Understanding historic building conservation

This book is part of a series on historic building conservation:

Understanding Historic Building Conservation
Edited by Michael Forsyth
9781405111720

Structures & Construction in Historic Building Conservation
Edited by Michael Forsyth
9781405111713

Materials & Skills for Historic Building Conservation
Edited by Michael Forsyth
9781405111706

Other books of interest:

Managing Built Heritage: the role of cultural significance
Derek Worthing & Stephen Bond
9781405119788

Conservation and Sustainability in Historic Cities
Dennis Rodwell
9781405126564

Building Pathology
Second Edition
David Watt
9781405161039

Architectural Conservation
Aylin Orbaşlı
9780632040254

Understanding historic building conservation

Edited by

Michael Forsyth

Department of Architecture and Civil Engineering
University of Bath

WILEY Blackwell

Contents

Preface

This is the first in a series of books that combine conservation philosophy in the built environment with knowledge of traditional materials, and structural and constructional conservation techniques and technology. The chapters are written by leading architects, structural engineers and related professionals and practitioners, who together reflect the interdisciplinary nature of conservation work.

While substantial publications exist on each of the subject areas – many by the present authors – few individuals and practices have ready access to all of these or the time to read them in detail. The aim of the series is to introduce each aspect of conservation and to provide concise, basic and up-to-date knowledge sufficient for the professional to appreciate the subject better and to know where to seek further help.

Of direct practical application in the field, the books are structured to take the reader through the process of historic building conservation, presenting a total sequence of the integrative teamwork involved. The second volume, *Structures & construction in historic building conservation*, traces the history of structures in various materials and contains much guidance on the survey, assessment and diagnosis of structures, the integration of building code requirements within the historic fabric and much else besides. *Materials & skills for historic building conservation*, the third volume in the series, describes the characteristics and process of decay of traditional materials which inform the selection of appropriate repair techniques.

The present volume, *Understanding historic building conservation*, discusses conservation philosophy and the importance of understanding the history of a building before making strategic decisions, the vital role of each conservation team member and the challenges of conservation at planning level in urban, industrial and rural contexts and in the conservation of designed landscapes. The framework of legislation and charters within which these operate is described; at the time of writing, designation legislation in the United Kingdom is due to undergo substantial reform over several years, and the context of this is comprehensively reviewed. The book provides guidance on writing conservation plans, explains the basic issues of costing and contracts for conservation, and highlights the importance of maintenance.

The series is particularly aimed at construction professionals – architects, surveyors, engineers – as well as postgraduate building conservation students and undergraduate architects and surveyors, as specialist or optional course reading. The series is also of value to other professional groups such as commissioning client bodies, managers and advisers, and interested individuals involved in house refurbishment or setting up a building preservation trust. While there is a focus on UK practice, most of the content is of relevance overseas (just as UK conservation courses attract many overseas students, for example from India, Greece, Australia and the USA).

Michael Forsyth

Acknowledgements

I am grateful to the following for permission to reproduce illustrations: English Heritage (4.1, 4.2, 5.2, 6.1–6.9); Mendip District Council (4.3); Eric Berry (5.1); Peter Gaskell (5.3); Bob Edwards (5.4, 5.5); Jeremy Lake (5.6, 5.7); John Winter (8.1, 8.2); Philip Whitbourn (10.2); Peter Davenport (13.1–13.4); Maintain our Heritage (16.1, 16.2).

Contributors

Timothy Cantell
Heritage and planning consultant, project coordinator of Maintain our Heritage and Manager of the South West Design Review Panel. Founding trustee of SAVE Britain's Heritage and was a trustee of Bath Preservation Trust. Formerly liaison officer for the Civic Trust and deputy secretary of the Royal Society of Arts. Interest in maintenance can be traced back to *Left to Rot* (SAVE, 1978) of which he was co-author.

Martin Cherry
Taught medieval history at the Universities of Exeter, St Andrews and Leicester, then moved into historic buildings conservation as fieldworker in Devon on the Accelerated Listing Resurvey. Worked on medieval houses in Kent for the Royal Commission on the Historic Monuments of England, then briefly as conservation officer in Leicestershire. Joined English Heritage, 1988; positions held include Head of Listing and Research Director. Publications on medieval and nineteenth-century architecture and conservation policy; edits *Vernacular Architecture*. Visiting Professor, Department of Architecture and Civil Engineering, University of Bath.

Nigel Dann
Building surveyor, senior lecturer and researcher, University of the West of England, Bristol. A founder and director of Maintain our Heritage. Management and conservation research projects for English Heritage, the Department of Trade and Industry, the Heritage Lottery Fund and other organisations. Extensive publications on conservation and maintenance issues and has co-authored (with D. Marshall) *The House Inspector* (Estates Gazette Books, London, 2005), a book on building defects and inspection.

Peter Davenport
Field archaeologist, particularly interested in the built environment, historic buildings and the relationship between urban archaeology, conservation and development. Studied, University of East Anglia and Keble College, Oxford, 1969–77. Archaeological excavations at Senior Supervisor and Deputy Director level, then appointed to Bath Archaeological Trust, 1980; Director of Excavations, 1990. Activities included excavation of the Temple Precinct under the Pump Room and, from 1996, research into the form and

function of the Roman town of Bath. Joined Oxford Archaeology, 2005, and Cotswold Archaeology, 2006, as Senior Project Manager. Considerable experience in the archaeological investigation of historic gardens. Long-term prehistoric to post-medieval research projects in southern Spain and Brittany, 1987–2002.

Geoff Evans

Quantity surveyor in private practice; senior partner, Bare, Leaning & Bare, specialising in historic buildings, conservation, repair and alteration. Projects from timber-framed buildings, vernacular dwellings, barns and chapels to major country houses, castles and, as named quantity surveyor, six cathedrals. Projects include: Dinmore Manor, Hereford; Ipswich Town Hall; the Castle, St Michael's Mount, Cornwall; major repair, maintenance and new works at St. Paul's, Canterbury, Wells, St. David's and Truro Cathedrals; Westminster Abbey; Abbey Gatehouse, Bristol; St Barnabas Church, Homerton; Roman Baths, Bath; Royal Foundation of St Katharine, East London; Knole, Sevenoaks, Kent (refurbishment of the private apartments); and the listed 1960s Blue Boar Quad, Christchurch, Oxford by Powell and Moya.

Keith Falconer

Head of Industrial Archaeology, English Heritage. Appointed survey officer to the Council for British Archaeology's Industrial Monuments Survey in 1971 and transferred to the Royal Commission on the Historic Monuments of England, 1981 then English Heritage, 1999. Coordinates industrial archaeological work across English Heritage from its Swindon base. Author of *Guide to England's Industrial Heritage* (Batsford, 1980) and co-author of *Swindon: the Legacy of a Railway Town* and numerous articles on the management of the industrial heritage. Visiting Fellow of the Centre for the History of Technology, University of Bath.

Michael Forsyth

Architect and director of the postgraduate degree course in the Conservation of Historic Buildings, University of Bath. Studied, University of Liverpool, held the Rome Scholarship in Architecture and, after residence in Italy, moved to Canada, working on the design of the new concert hall for the Toronto Symphony Orchestra with the architect Arthur Erickson. Lectured at University of Bristol, 1979–89 and has lived and practised in Bath since 1987. Books include *Bath – Pevsner Architectural Guides* (Yale University Press, 2003) and *Buildings for Music: The Architect, the Musician, and the Listener from the Seventeenth Century to the Present Day* (MIT Press and Cambridge University Press, 1985), which won the 19th Annual ASCAP–Deems Taylor Award. Translations in French, German, Italian and Japanese. Holds a Doctor of Philosophy degree of the University of Bristol.

James Maitland Gard'ner

Studied architecture at the Victoria University of Wellington, New Zealand; completed training in building conservation at the Architectural

Association, London. Joined English Heritage as a historic buildings architect, advising public and private sector owners on the care of historic buildings and working on the conservation and adaptive reuse of English Heritage's properties in care. Involvement in conservation planning included the preparation of plans for English Heritage's own sites and organising seminars. Now resident in Australia.

Colin Johns

Architect and planner; member of the Institute of Historic Building Conservation. Sole practitioner in Bradford on Avon, Wiltshire, specialising in the conservation of historic buildings including individual projects, conservation area assessments and community action. Visiting lecturer in conservation legislation and practice at the Universities of Bath and Plymouth. Chairman of the United Kingdom Association of Preservation Trusts and architect to the Wiltshire Historic Buildings Trust Ltd.

Jeremy Lake

Worked with the National Trust, publishing the book *An Introduction and Guide to Historic Farm Buildings* (National Trust, 1989) and as a field worker on the Accelerated Listing Resurvey of England. Joined English Heritage, 1988, carrying out a range of surveys in urban and rural areas, and, since the mid-1990s, thematic listing surveys on chapels, military and industrial sites and farmsteads that have connected designation to guidelines for reuse and management. Since 2002 has worked with Characterisation Team, developing new methods for understanding and managing the historic environment, including landscapes and rural buildings. Extensive publications on farmsteads, military sites and chapels.

Jonathan Lovie

Holds degrees from the University of St Andrews, including Master of Philosophy. From 1994, developed a consultancy undertaking historic landscape research and conservation management plans for private clients and public bodies. Consultant Register Inspector, English Heritage, 1998–2004, with particular responsibility for revising and upgrading the *Register of Parks and Gardens of Special Historic Interest* in the south-west of England. Involved in thematic studies leading to the inclusion of significant numbers of public parks, cemeteries, institutional and post-war designed landscapes on the *Register*. From 2004, additional to ongoing private practice, appointed as Principal Conservation Officer and Policy Adviser in England to The Garden History Society. Lectures on courses relating to historic landscape research and conservation.

Duncan McCallum

Policy Director for English Heritage. Worked for Cumbria, Durham and Devon County Councils in various historic environment roles, then joined English Heritage, the government's adviser on the historic environment, in 1996. Produced England's first annual state of the historic environment report (now known as *Heritage Counts*). Became Head of Planning and Regeneration, then Policy Director. Holds an honours degree in Town and Country Planning and a Master of Arts degree in Conservation Studies from the Institute of Advanced Architectural Studies, York, and is a member of the Institute of Historic Building Conservation.

Martin Robertson

Historic buildings consultant. Educated at Dulwich College, Downing College, Cambridge (History Tripos) and University of Edinburgh (History of Art). With the government historic buildings service from 1968 working for two ministries, then English Heritage, becoming Principal Inspector of Ancient Monuments and Historic Buildings and leader of the East Midlands Team. Gained wide experience in historic building types, their problems, conservation, grant aid, taxation and legislation. Contributed to the advice given in PPG15 (1994). Private practice from 1993 working for private clients and national and local bodies including Cadw. This work has given insights into conservation from the owner's perspective, the constraints placed on people and the occasionally bizarre results. Lecturing includes University of Bath.

Adrian Stenning

Chartered quantity surveyor and senior partner, Bare, Leaning & Bare, specialising in the repair and conservation of, and sympathetic extension to, historic buildings, from formal classical architecture to vernacular timber-framed buildings. Member, RICS Register of Surveyors Accredited as experienced in Building Conservation Work. Member of the Association for Studies in the Conservation of Historic Buildings (ASCHB). Lectures on quantity surveying with reference to historic buildings at the Universities of York and Bath, and for the Society for the Protection of Ancient Buildings (SPAB) as part of their twice-yearly repair courses.

David H. Tomback

Development Economics Director, English Heritage. Chartered surveyor, 1974; worked with surveying practices and a development company, then as an American bank's in-house property adviser. Helped run a Swiss-owned property development company, 1987–91. Formed own practice; joined English Heritage, 1993, advising on in-house commercial property, grant levels, enabling development, reuse of redundant historic buildings, and conservation economics. Involved in economic studies: *The Investment Performance of Listed Buildings, The Listing of Buildings – the effect on value* and *The Value of Conservation.* Responsible for *Heritage Works – The use of historic buildings in regeneration* (2006). Member, working group producing English Heritage's *Policy Statement and Practical Guide on Enabling Development,* National Audit Office panel of expert valuers on the disposal of NHS Estates properties and RICS Public Sector Advisory Committee; joint chair, NHS Estates/English Heritage working party. Visiting lecturer, College of Estate Management and universities of Bath, Bristol and Oxford Brookes.

Giles Waterfield

Independent curator and writer, Director of Royal Collection Studies and an Associate Lecturer at the Courtauld Institute of Art. He is Chairman of the Trustees of the Charleston Trust, and of the Paul Mellon Centre for Studies in British Art, and was Director of Dulwich Picture Gallery, 1979–96. He was joint curator of the exhibitions Art Treasures of England at the Royal Academy of Arts in 1998, In Celebration: The Art of the Country House at the Tate Gallery in 2000 and Below Stairs at the National Portrait Galleries in London and Edinburgh, 2003–2004. Publications include *Palaces of Art*, *Art for the People* and *Soane and Death*. He has published three novels, *The Long Afternoon*, *The Hound in the Left Hand Corner* and *Markham Thorpe*.

Philip Whitbourn

Trained in Architecture and Town Planning at University College, London and spent ten years in architectural practice before joining the Historic Buildings Division of the former Greater London Council in the mid-1960s. After 20 years with the GLC, became chief architect to English Heritage for some ten years. Served as Secretary, ICOMOS-UK (The International Council on Monuments and Sites-UK), 1995–2002. Elected Fellow of the Society of Antiquities, 1984; awarded the OBE., 1993. Holds a Doctorate in Town Planning.

John Winter

Completed studies, Architectural Association, London, 1953; elected Associate, Royal Institute of British Architects. Completed first building, 1956, a house included in Pevsner's *North-East Norfolk*. Studied, Yale University, 1956–7 then worked with Skidmore, Owings and Merrill, San Francisco and Ernö Goldfinger, London. Taught, Architectural Association, London, 1960–64 and formed John Winter and Associates, 1964. His buildings have been extensively published in England, Europe, North America and Japan and received numerous awards. Awarded MBE for services to architecture, 1984. Member, Royal Fine Art Commission, 1986–95; Council of The Architectural Association, 1989–95. Appointed trustee of the Architectural Association Foundation, 1995 and of DoCoMoMo, 1999 and as Architectural Adviser to the Heritage Lottery Fund, 1996. Author of *Modern Architecture* (Paul Hamlyn, 1969), *Industrial Architecture* (Studio Vista, 1970) and of two co-authored books.

1 The past in the future

Michael Forsyth

Buildings can be victims of conservation interests. An Australian engineer, Tony Graham, bought the last remaining ironworks near Mells in Somerset. He planned to convert the handsome but decayed office building into a house. Different conservation bodies then descended. The site contained greater horseshoe bats, and became a Site of Special Scientific Interest and could not be disturbed. The Victorian Society, on the other hand, had the site listed and demanded that the office building be restored. The industrial archaeologists, meanwhile, took an interest in the foundry ruins and declared that the site must be cleared. Naturalists discovered rare ferns and said that the site was not to be touched. After prolonged disagreement the owner, wanting simply to proceed with the work, requested a site meeting with the local council and the parties involved in order to resolve the situation. Meanwhile, some boys caught in a rainstorm sheltered in the building and lit a fire to dry their clothes. The building caught fire and burned down.[1]

In the United Kingdom half of the building industry's workload, including maintenance, is concerned with existing buildings. Yet conventional training for architects and engineers provides little or no guidance on the care of existing buildings and too many historic structures are still being damaged by unsympathetic treatment. Despite this, and despite the changed construction methods and materials that replaced building techniques lost during the twentieth century, traditional craft skills are steadily being rediscovered. This is due in no small part to the series of fires at York Minster in 1984, Hampton Court, Surrey, in 1985, Uppark, West Sussex, in 1989 and Windsor Castle in 1992. Meanwhile, since the mid-1970s we have swung from an era that saw destruction of historic town centres and country houses alike, to a planning ethos where 'heritage' and 'conservation' are words that recur. We border dangerously on a museum mentality that fiercely resists change.

The Venice Charter – the philosophical manifesto produced by the International Congress for Conservation in Venice in 1964 – defined several possible approaches to conservation. Preservation involves the minimal repair and maintenance of remains in their existing state. Restoration involves the removal of accretions to return a building to an earlier state. Reconstruction also involves returning a building to an earlier state, but involves introducing new – or old – materials to the fabric. Conservation may involve one or more of these, as well as the adaptation of buildings to new uses. Historically, the stance that we have taken on building preservation has constantly shifted, and the only certainty is that tomorrow's conservation philosophy will be different from that of today.

Until William Morris founded the Society for the Protection of Ancient Buildings (SPAB) in 1877, a ruthless philosophy of restoration and reconstruction was normal. The usual approach to church restoration was to undertake whenever possible a radical return to a definite style and to make the building look smooth and crisp and symmetrical like the new churches of the Gothic Revival. The eighteenth-century restorations of James Wyatt and his contemporaries posed a greater threat to medieval buildings than either neglect or fire. Wyatt's new west front to Hereford Cathedral of 1788 provoked an outcry even at the time. In 1818 at Chester Cathedral, Thomas Harrison added squat corner turrets to the south end of the transept. Anthony Salvin (a pupil of John Nash of Regent's Park fame) in 1830 refaced the south transept of Norwich Cathedral, replacing the original Perpendicular with a Norman design to match the north transept. At Canterbury in 1834 George Austin demolished the Romanesque north-west tower and replaced it with a copy of the south-west tower for symmetry. In the 1830s, the thirteenth-century nave of Southwark Cathedral was demolished, and at Bath Abbey a programme of correcting the building, including the addition of false flying buttresses, was carried out by George Phillips Manners. In 1870 Scott demolished the whole east end of Christ Church Cathedral, and rebuilt it in Norman style. And so the list goes on.

In the past, different categories of buildings were thought worth preserving at different times – mainly because they reached an age at which they were regarded as venerable. By the late nineteenth century, medieval buildings were sufficiently esteemed to be preserved for their antiquity. The first protective legislation was the Ancient Monuments Act 1912, which served to preserve decayed and obsolete structures that had artistic or historic interest. By the early twentieth century Jacobean and Queen Anne buildings became respected, but later Georgian buildings only gained sufficient historical perspective to be regarded as worthy of protection with the formation of the Georgian Group in the 1930s. The turn of Victorian architecture came much later. The 1960s and 1970s are now recognised as historical eras in their own right, and eminent listed buildings from this era now include London's Centrepoint office block and Norman Foster's high-tech Willis Faber & Dumas building, Ipswich.

With the 1944 Town and Country Planning Act, historical buildings were first seen for their townscape value as groups rather than on their own architectural merit. But the conservation movement as we know it was slow to gather pace following this basic legislation. Widespread destruction in the Second World War, and the social optimism of the era that followed, led to a comprehensive attitude towards redevelopment. In a lecture given at Bristol University in 1947 and published in his collection of essays *Heavenly Mansions*, Sir John Summerson pleads for the preservation of outstanding historic buildings. But his list of 'types of buildings which may in certain circumstances deserve protection' reads from our perspective as positively advocating the comprehensive redevelopment schemes that swept away the centres of most historic English towns and cities after 1945. Following Sir Patrick Abercrombie's post-war plan for Bath, which proposed that the Royal Crescent be converted into council offices linked to a modern block at the rear, about one third of Bath's historic city – about 1000 Georgian buildings, of which some 350 were listed – were demolished between 1950 and 1973. By the 1970s, traffic problems added to inner-city congestion and decay. The countryside also suffered as badly. Multiple death duties during the First World War, often within months, caused the downward slide of hundreds of country houses. In the period from 1945 to 1973, 750 major country houses were demolished, and the impossibility of their upkeep culminated in the Labour government's wealth tax of April 1974 when the top rate of tax increased from 90% to 98%.

But the tide was turning. The Civic Amenities Act 1967 called for local authorities to designate conservation areas. Conservation studies were published in 1969 for Bath, Chester, York and Chichester[2] to examine methods of funding and repair of historic buildings. In 1973 an influential book, *The Sack of Bath* by Adam Fergusson, published for the first time the scale of destruction in this most intact of historic cities. Marcus Binney created in 1974 The Destruction of the Country House exhibition at the Victoria and Albert Museum, showing grim pictures of architectural decay and demolition. In the same year he set up the campaigning organisation SAVE Britain's Heritage, and the following year was European Architectural Heritage year. In 1976, faced with a flood of country houses coming onto the market, the Labour government replaced the wealth tax with a new Finance Act. Moreover, the new affluence of the 1960s brought about the car-owning society – by 1964, 20 million private vehicles were on the road – and this caused a new interest in the countryside.

From the early 1970s through to the Thatcher years of the 1980s, vast numbers of city dwellers dreamed of moving to the countryside and bought period cottages as first or second homes. Country house visiting became a major pastime and membership of the National Trust soared, doubling to 550 000 between 1972 and 1975, and reaching 850 000 by 1980. Particular interest in visiting historic gardens resulted, in the late 1990s and early twenty-first century, in members of the Historic House Owners' Association (HHA) rebranding their houses, open to the public, as gardens with houses attached rather than historic houses with gardens. In cities, too, fuelled by the country house interiors style, upstairs-downstairs films and the desire to own a period home, there was everywhere the wish to preserve or evoke the past. The heritage society had arrived.

With this swing of the pendulum came the new danger that our historic cities would lose their vitality and become heritage museums. There is a tension between keeping cities alive and conserving their historic fabric, a dilemma between 'development' and 'conservation'. Conservation has as much to do with breathing new life into old buildings as it has with repair. Nearly all buildings have evolved over their lifetime, adapting to the needs and uses of successive generations. Buildings decay when they are abandoned without a use, and their spirit dies when they become frozen in time as near museum pieces. Historically, buildings that lost their purpose disappeared, and those old buildings that are still with us have usually undergone frequent adaptation or changes of use. When buildings have a viable use, there is the incentive to repair and maintain the fabric, while old buildings deteriorate rapidly when neglected or empty. Urban regeneration is a vital ingredient in conservation, involving a partnership of business initiative with the skills of town planning and heritage management. Buildings should preferably maintain their original purpose, but the door should always be open where appropriate to new uses, adaptability and extension. The conversion of redundant warehouse buildings has revived many dockland areas. The reuse of St Katharine's Dock in London, built in 1827–29, as apartments and a hotel led to numerous other schemes, including the conversion of Jesse Hartley's Albert Dock, Liverpool, of 1839–45 into a recreational and residential area. The conversions into art galleries of the Castellveccio in Verona by Carlo Scarpa and of a redundant Paris railway station at the Musée d'Orsay are outstanding European examples.

Another important field for conservation at the level of urban planning is the consideration of new buildings within historic cities. An interesting example of the possible scope of this is the Historic Royal Palaces Tower Environs Scheme. Under the scheme, sightlines from within the Tower of London were projected into infinity to define the maximum height of new buildings around the Tower. This ensures that no building in the City or beyond may be visible from the enclosure of the historic buildings.

If one end of the conservation spectrum embraces the urban management of entire towns and cities, the other end, involving the care of individual buildings, ultimately concerns good construction practice and an understanding of how buildings were originally designed. At least when working on eighteenth- and nineteenth-century buildings, the conservation architect requires knowledge of classical architecture, in addition to a philosophical standpoint and knowledge of traditional materials. Western industrial cities – whether London, Paris or New York – can be thought of as fundamentally classical. Each comprises a legacy of buildings, whether classical, Gothic or whatever, that were originally designed by architects trained in the classical tradition. Builders, too, had knowledge of the same visual language, and from the eighteenth century onwards speculative houses were built with the aid of pattern books, such as Battey Langley's *Builder's Jewel* of 1739. These well-thumbed, pocket-size books explained everything the builder needed to know, from the construction of classical orders to the geometry of mouldings and the proportions of a room. Sadly, the classical training – with students routinely producing astonishingly competent renderings – died out in schools of architecture in the early 1950s. But when working on historic buildings, it is essential for present-day architects to have a working knowledge of those same principles in order to design even a glazing bar or a balustrade or to position a dado rail.

Before undertaking any conservation work on a building, it is essential to understand the building by carrying out a careful assessment of its history, the decay of its fabric and the causes. Repair work should always respect the history of a building, and this appraisal will help to keep intervention, repair and treatment works to a minimum. For any historic building this will involve an archival investigation and a survey of the building structure and fabric. It is then possible to make a conservation plan that assesses what needs to be done – if anything – and the repair techniques and technologies that will be used. If the planner and heritage manager are significant in conservation initiatives at an urban level, then individual building repair increasingly involves the architectural historian and building archaeologist, in addition to the team of architect, engineer, quantity surveyor and builder.

Every building, however humble, possesses a history, and buildings from different periods and regions are unique. All historic buildings undergo cycles of alteration in their lifetime. Typically, minor repairs are carried out periodically, with programmes of major maintenance, renovation and modification taking place at less frequent intervals. This pattern may alternate with periods of relative inactivity and perhaps neglect. Major changes are usually made to buildings to modify or extend their use, to update their style, and particularly to repair fire damage. Most country houses have suffered fires, while theatres in the western world, before modern fire prevention codes were developed in the second half of the nineteenth century, suffered major fire damage on average every eighteen years.

The first task is to carry out investigations through a combination of archival research and on-site survey. For the archival research, county and city archives, local libraries and the National Monuments Record in Swindon are usually invaluable sources. Where appropriate, an architectural historian may carry out this work. Meanwhile, engineering and other investigations into the building fabric, tailored to each situation, should be carried out. These will reveal how the building stands, and whether or not any structural work is necessary. It is vital that all members of the team understand the building and that a sequence for the work is planned. A shortfall in knowledge leads to surprises, and buildings are most at risk when they are being worked on. During investigations, appropriate caution and a basic knowledge of historic building technology are necessary. One builder took up all the floorboards in a Georgian house to examine the joists, not realising that the floors act as plate membranes, and the house collapsed.

With this information, it is then possible to assess the building and form a conservation plan. This document sets out the architectural history, and then presents a rationale and policy for the proposed works. The architect has to decide how far to wind back the clock and, in particular, a view has to be reached on the dilemma between respecting the intentions of the original architect and respecting the history of the building. In Bath, the Victorians lowered the sills of most Georgian houses and inserted plate glass into new, heavier sash frames with horns, in place of the original sashes with glazing bars and thin meeting rails. The question arose as to whether the sills in the Royal Crescent should be restored to their original height, at the cost of internal damage and disruption. If not, should Georgian glazing bars be inserted into the enlarged windows, which were never intended to be subdivided and where suitable proportions might not be possible? Several years ago, the Bath Preservation Trust raised the sills of its headquarters at 1 Royal Crescent, but a recent debate with English Heritage about the remainder of the crescent decided against alteration.

Views constantly change, and current thinking leans towards respecting the history of a building. When English Heritage restored Lord Burlington's Chiswick House, the late eighteenth-century wings that Henry Holland added to this freestanding Palladian villa were removed. This undoubtedly enhanced the original building, but it is unlikely that the demolition would have happened under today's conservation philosophy. Taken to extremes, peeling away layers of history may leave alarmingly little. On the Acropolis in Athens, accretions of Byzantine and medieval additions were radically demolished to reveal the fifth-century BC buildings – the Parthenon, Propylaia and Erechtheum – but all that remained were ruins.

Concurrently with making these decisions, the architect must also plan the work to satisfy present requirements for function and safety in a way that is compatible with the building. The protection of life is paramount, but it is arguable that safety legislation for buildings exceeds that for other everyday situations, such as underground railway platforms and roads. There is an inherent conflict between conservation legislation and building regulations, and there are many situations that current codes of practice or conventional methods cannot deal with in historic buildings, at least not without causing unacceptable damage. Fire engineering and floor loading are just two among many areas where creative solutions – or lateral thinking – can be used to provide acceptable alternatives for a more sympathetic treatment. For example, if a building is being converted to office use, heavy storage may be placed in the basement instead of floors being invasively strengthened to recommended levels. A lintel in an old building, even when badly distorted, may be left undisturbed if it is still performing. If a timber beam works despite signs of decay or deformation, it may not require additional work, while deflection may not be a problem if flexible finishes are used. Sprinklers, even in domestic situations, may be acceptable and less invasive than partitioning and fire doors. Sometimes a sceptical approach is necessary. The fact that a building has stood for 200 years may be eloquent proof of structural soundness despite rulebook calculations that show its structure to be inadequate.

The next step before undertaking repair work is to identify which techniques are appropriate and decide how far one goes. Experience of traditional construction and skills is necessary, together with knowledge of the characteristics of materials, including how they decay and the reasons why modern materials frequently cause damage to old buildings. For example, lime-based products are fundamental to conservation work because they are flexible and breathable. Mortar must be softer than adjacent masonry to absorb movement and to be 'sacrificial' to the original stone or brickwork. Because of chemical reaction, cracks in lime mortar are self-healing, while hard, impermeable Portland cement mortar traps moisture and quickly loosens with freeze–thaw action.

When planning the works, there are also several philosophical principles to follow. The first is *minimal intervention*. The current philosophy is that the total fabric and structure of historic buildings, not merely surface appearance, is integral to their character. There should be minimal interference with, or damage to, the original structural fabric. It is also important where possible to avoid change to the original structural mode of behaviour. *Reversibility* is also a keyword, and repairs should be capable of being undone in the future, as increasingly compatible materials and techniques are developed and conservation philosophy evolves. The only certainty is that future generations will regard what we carry out today with scepticism. Many twentieth-century conservation techniques have led to problems that cannot be reversed. It is also good practice to *conserve as found*. If the footings of a medieval timber-frame barn have settled, it is unwise to jack up the structure from its new equilibrium to the original alignment. Repairs should be *like for like*, using either original or compatible materials. In the twentieth century, many unsuitable methods were used in repair work. Iron and steel rods were commonly inserted into stone as reinforcement and grouted with Portland cement, but the iron rods soon rust and expand, cracking the masonry. Iron and steel were superseded by stainless steel, but this too cracks masonry, with its different coefficient of expansion.

These principles are only a guide, and traditional materials and repair methods are not always best. With historic roof structures, inserting steel to repair rotting members, rather than carrying out timber repairs, may avoid sections of the historic fabric being cut out and lost – and may be more reversible. Alternatively, a steel flitch plate inserted into a timber beam may be the least visible type of repair, but at the loss of reversibility. It is also unrealistic – and pointless – to search doggedly for authenticity in the use of materials. We would not wish to use toxic – and illegal – lead paint except in buildings of outstanding importance, nor to paint the facades of a listed weekend country cottage with ox blood. Nor could we live with truly authentic interiors. A glance at the squalid conditions suffered by our forefathers, as recreated at Cardiff's Museum of Welsh Life, would quickly dispel such thoughts. And for grander houses, the popular historic ranges of paint colours are in fact muted versions of the original colour schemes, which we would find hard to live with today.

In the present ultra-conservationist climate, it is arguable that conservation legislation has gone too far in certain situations. Conservation laws can have the opposite effect to what was intended, and some relaxation of guidelines would sometimes better serve the interest of a building. The blanket refusal to allow any change is artificial and can ultimately be damaging to a building. Throughout history, buildings have adapted to changing needs and situations. Sometimes a local authority's refusal to grant listed building consent may be an excuse for doing nothing. Eighteenth-century townhouses, with strictly regulated street facades, were freely extended and altered at the rear to allow for changing requirements – the so-called Queen Anne front and Sally Anne back.

Conservators today approach all these problems as a doctor to a patient, where radical surgery is not a preferred option; perhaps a drug, an aspirin or pacemaker can keep the problem at bay. All engineering and other repairs are invasive, but the conservation professional will usually strive for localised repairs. Conservation involves taking what is there and improving it before cutting or adding, while at the same time being able to show that present-day standards of public safety and comfort and important legislative requirements are satisfied.

Endnotes

1 Christopher Booker's Notebook, *Daily Telegraph*, 15 July 2001.
2 Ministry of Housing and Local Government, *A Study in Conservation*, 4 vols (HMSO, London, 1968).

2 Architectural history and conservation

Martin Cherry

Introduction

Most people, historians among them, have some attachment to the past. But historians do not necessarily seek, according to some universal rule, or by temperament or vocation, to preserve or return to it. While many historians have been active in the conservation movement, many more have not, and there is a perception among those responsible for the management of the historic environment that academic architectural history has remained detached from the conservation front line. This chapter attempts to tease out some of the issues behind that perception by taking a historical perspective. It is not possible in the space available to analyse the full complexity of the discourse between architectural history and conservation. Instead, a small number of historic moments or trends are selected that may go some way to explain where we are and how we got here. The main impression is that, despite what one might expect, the two have not made natural, or even comfortable, bedfellows.

To a certain extent, the relationship between architectural history and conservation parallels an old debate within the archaeological profession. Conservation employs the majority of archaeologists in Britain. This is probably the case with architectural and building historians, too. Most archaeology in Britain focuses on the long-term needs of surviving deposits, sites and artefacts: research supports environmental management. Archaeologists are conscious that to walk away from an excavated site and leave it an empty shell would provoke public criticism. Driven by the fact that excavation damages or even destroys sites, non-intrusive analysis is preferred and conservation is integral to modern investigative approaches.[1] Academic archaeology is seen increasingly by this growing phalanx of archaeological practitioners as remote and irrelevant.

Transposing this observation to the relationship between academic history and conservation presents both insights and pitfalls. First, historical research and building analysis, unlike excavation, do not in themselves cause damage, and one of the prime drivers of archaeological policy – to protect the resource from the effects of excavation-led analysis – does not appear, on the surface at least, to apply. But the assessment of significance based on historical research can have a dramatic impact on how a building is passed on to future generations. Some outcomes can be radically intrusive, such as the removal from a building of accretions that are judged to belong to an insignificant period of the building's history (as once was quite common). At the other extreme, changes can be so subtle and sensitive that they are barely noticeable and cannot easily be read as part of the building's story.[2]

But the live issue is less about the philosophical implications of intervention than about mundane but critical matters such as research agendas and the availability of skills and training. There is a view that the key research areas that are needed to help secure the future of the historic environment are not being fully developed in the university sector; nor is architectural history being reflected adequately even in applied conservation courses.[3] This may partly be explained by mistrust within academe about practising history in the public interest. More likely, it is simply the failure until recently to address the mismatch. This became an issue of public policy in 2001 when the government encouraged English Heritage to liaise much more closely with the universities and other organisations in the development of research programmes. A major stepping stone was reached when the Arts and Humanities Research Council (AHRC) upgraded the study of the historic environment to become a fundable subject area in its own right. Another was the publication of English Heritage's first systematic research strategy in 2005.[4]

Sound conservation depends on the availability of accurate historic data authoritatively interpreted. Especially with complex buildings, this requires skill sets (in recording and analysis) that are not well developed in the university sector. This apparent lack may reflect a greater emphasis being placed in architectural history courses on documentation, literary sources and theoretical underpinning than on close fabric analysis. Such generalisations are difficult to quantify but bibliographical studies have demonstrated reluctance on the part of general historians to use architectural evidence.[5] Interestingly, the opposite situation – the failure of building analysts to exploit multidisciplinary document-based approaches – has also been quantified.[6]

More sinisterly, the lack of analytical skills has been attributed to major methodological shifts within the universities in recent years. A leading architectural historian, Frank Salmon, asserts astringently that 'much writing about historic architecture by younger writers . . . is couched in the complex language of post-modern discourse, but leaves one without the confidence that authors have either the vocabulary or the visual and analytical skills necessary to describe the formal appearance or material existence of a building'.[7] While the relatively high level of good-quality published architectural history belies this view if extended to the discipline as a whole, it is true that organisations tasked with managing the historic environment find it difficult to recruit people with the hands-on skills of detailed fabric and documentary analysis and look increasingly to archaeologists and others in the private sector.

But the debate is about more than academic attitudes and the availability of finely honed forensic skills. Few would dissent from the view that it is essential to have access to reliable information expertly interpreted in order to decide how best to conserve, manage or alter a historic site. Assuring the credibility of specialist views is also essential at a political level, as part of the planning and consent process, in inquiries and courts of law and, increasingly, in order to retain public confidence and support. It is in this latter area – what might broadly be called public engagement – that things are changing fastest and confronting historians and conservation practitioners alike with new challenges.

Increasingly over the last 150 years, expert opinion has influenced attitudes to architecture. It would be difficult prior, say, to the 1930s to characterise this elite as being made up wholly of professional or academic historians, but they were historians nonetheless, more than antiquaries, and they possessed a clear picture of how the historic environment should look. Their views have helped determine what is considered excellent or worthwhile and what should be rejected as dross or insignificant. This has inevitably influenced decisions as to what should be allowed to survive. As elitist views changed, so did public policy. The receptiveness of government to such views grew from the time of the first listing legislation in the 1940s and may be said to have peaked with the publication of Planning Policy Guidance Note 15 (PPG15) in 1994, which formed something of a high-water mark for conservation. During the 1960s and 1970s, as a wider public became involved in the politics of conservation and historians explored new social dimensions and the dynamics of community identities, so specialists, by no means all based in universities, responded and researched wider ranges of building types and historic neighbourhoods that in turn came to be seen as worthy of protection.

The situation at the dawn of the twenty-first century is somewhat paradoxical. The electronic age has sharpened the tendency, visible in the cultural radicalism of the 1960s, to treat specialist views sceptically. Neatly put by the columnist Richard Tomkins, these changes have resulted in 'the debunking of expertise and demystifying of talent. Today, authority figures of all kinds, be they presidents, scientists [we may add professional historians and conservationists] or art critics, are less respected and less trusted than they once were.'[8] These are some of the issues this chapter sets out to examine.

The paradox is that whereas the stimulus to involve professional historians was once primarily in order to strengthen the case for remedial and sustainable intervention in a historic building or site through better understanding, historic research has subsequently been employed for a greater range of purposes. It still has to help make the case that a 'heritage asset' has historic significance. But it has not only to do this in the face of increasing scepticism about the integrity and credibility of specialist views but also to offset the competing values and demands of local communities and other stakeholders. Conservation plans, for instance, now increasingly a requirement for both public funding and planning consent, require specialist heritage views to be placed alongside other sets of values, some of which may well be competing. The implications of all this for current policy and practice are examined in the final section of this chapter.

The role of architectural history in the management of historic buildings over time is a complex one, and current practice has its roots in a number of traditions. Much ink could be spent over when architectural history emerged as a distinct study area. Like conservation itself, it has what might be termed a 'pre-history'.[9] Modern conservation may be said to have emerged when it became motivated less by overt political or religious aspirations and more by a sense that a historic building had intrinsic value as a source of evidence for the past, a document, as it were, whose rarity and vulnerability could be measured as well as its significance. Architectural history may be distinguished from antiquarianism when it begins to work within a rigorous methodological framework and also when it, too, becomes consciously freestanding from religious motives. It is more than simply a classification of parts or periods, orders or styles in the manner of Rickman's categorisation of English Gothic into Early English, Decorated and so forth, important though this was. David Watkin was the first to see the significance of Edward Freeman (1823–92) in this respect.[10] His *History of Architecture* (1849) was a reaction against what he considered to be the hijacking of architecture by the ecclesiologists, High-Church Anglicans who equated certain matters of style and planning with true Christianity. His aim was to produce a history where the subject was 'not reduced to matters of antiquarian or ecclesiastical research'; to foster the study of architecture 'in its proper position as a branch of mental philosophy'; and to make 'an attempt at a philosophical history of the science of architecture'.[11]

This aspiration to inject greater rigour and 'science' into architectural history is one of several strands evident in the middle years of the nineteenth century that are relevant to an understanding of the discipline's role in modern conservation. It is worth briefly mentioning some others. Freeman's history was of medieval architecture, mainly because that was where his interest lay, and partly because, like many of his day, he saw it as marking the height of English achievement. He saw in the very last years of the Perpendicular a period of decay, and struggled with, but was honest enough to admire, the greatest Renaissance architects, Brunelleschi especially. His history was the work of a young man and he might have written differently in later life. For in his great history of the Norman Conquest, his notions of 'philosophical science' emerge more clearly. These stress the continuities of English society and institutions that weathered the conquest and led cumulatively to the distinctive constitution of his own time. If seen in terms of improvement, this becomes a Whig interpretation of history. But it also allowed for the values of a historic epoch to be taken on their own terms. In this respect, his slightly older contemporary George Aycliffe Poole (1809–93) was more perceptive. In his almost exactly contemporary history of architecture, Poole admits the transition to Perpendicular to have been one of the two most interesting episodes in English art.[12] 'Interesting' rather than 'outstanding' or 'worthy' is the key word, and in a telling passage about builders and patrons he explains that intrinsic interest '. . . is not *our* estimate of their character, or *our* way of expressing it, but it is *their* estimate, and *their* way of expressing [his italics]' that matter and lead to a full understanding of their work.[13]

To the two attributes evident (albeit in tentative form) in the writing of Freeman and Poole – seeking a rigorous methodology (a 'science of architecture') and looking at the work of historic architects and buildings objectively in their own terms – may be added a third. In 1862 the architectural historian James Fergusson (1808–86) produced an international account of architecture that brought the story entirely up to date: a *history* of architectural styles from the Sistine Chapel to the Houses of Parliament.[14] The fact that Fergusson was using architectural history as a means to counter what he saw as the evil of 'copyism' in modern architecture does not reduce the importance of his work in extending the remit of architectural history to the present day. The English Heritage post-war listing programme of the 1990s, to which we return later, was similarly designed to raise awareness of the quality of modern architecture through research: it had redoubtable antecedents.

That some of the prerequisites of modern architectural history were in place well before the launch of the Society for the Protection of Ancient Buildings (SPAB) in 1877 does not mean that architectural historians necessarily favoured preservation. Robert Willis (1800–75), philosophy professor, founder member of the British Archaeological Association and co-author of one of the 'pioneering work[s] of rationalist architectural history', a detailed four-volume account of the buildings of Cambridge University, appears at times to have been indifferent to the fate of historic fabric.[15] He assumed that much of what he so minutely recorded would be swept away in time the better to meet modern needs, and he did not object to Waterhouse's plans for extensive demolition at Pembroke College that in the event resulted in the architect's dismissal.

Architects, historians and popular perceptions

Even William Morris and the SPAB seldom sought to intervene when threats occurred to private houses, so deeply engrained was their respect for private property. Yet the idea that there was a public benefit attached to historic buildings lay at the heart of SPAB thinking and some felt that the nation could claim moral ownership of them.[16] It was to take many decades before that view became more widely held. The precursor body to the Survey of London was at first an overtly preservationist movement, established in 1894 *for* the people of London, and the Royal Commission on Historic Monuments, established in 1908, also set out to illustrate the 'culture, civilisation and conditions of life of the people in England'.[17] The National Trust was similarly sweeping in its aspirations. But the discourse was dominated by paternalistic attitudes, values of significance being determined on behalf of, rather than by, the people. The tone is well expressed by the architect Arthur Stratton, writing in 1910: 'Love of home is a strong characteristic of the English race, yet as a nation England has done little to preserve what has been bequeathed from the past.'[18] This view of what was valuable in the English historic landscape could easily slip into a nostalgia for Old England and is reflected in the large genre of books about picturesque old towns and villages and the ancient countryside, many published by Batsford, which often consciously rejected academic modes of thought and were sometimes anti-democratic.[19]

Architects and architectural journalists, especially in the pages of the *Architectural Review*, kept a vigorous debate alive in the interwar years and were critical in fashioning (or helping to articulate) public opinion during the seismic shifts of the 1960s when conservation became a force in the land. Chris Miele neatly characterises the significant changes in elite views in the post-war period by contrasting three influential documents.[20] The survey by Holden and Holford of wartime destruction in London stressed the continuities of historic architecture. Summerson saw Georgian London as a model for rebuilding sane and human townscape. Both products of the 1940s, these works combined shocked reaction to calamitous change inflicted by enemy action with optimism coloured by modernism. By the late 1960s and early 1970s the tone was one of outrage against the wantonness of developers and planners as reflected in Hermione Hobhouse's polemic, *Lost London* – clearly 'protest literature', but highly informed. The classic example of the latter was Ian Nairn's *Architectural Review* issue entitled *Outrage*, and this period has subsequently been dubbed 'the heroic period of conservation'.[21]

What was the role of architectural history and historians in the events so summarily set out in the previous paragraph? Not everyone would subscribe to David Watkin's view that the intellectual discipline of art – and architectural history in its present form – was established in Britain and America only after the influx of art historians fleeing Nazi Germany in the 1930s. According to Watkin, the arrival of people such as Pevsner, Gombrich and Wittkower injected a standard of scholarship that made the 'comfortable English world' of 'gentlemen . . . from privileged backgrounds' for whom 'historical research was rarely the central activity of their careers' look 'extremely amateurish'.[22] This is overstatement to make a point, but it is true that even in 1956, when the Society of Architectural Historians of Great Britain (SAHGB) was founded, it was by no means self-evident that such a discipline existed and 'at that time there were few persons who could be considered architectural historians professionally'.[23] It seems an arid debate to question whether or not the likes of Garner and Stratton, Gotch or Ward, who made such major contributions to our knowledge of Renaissance architecture, were architectural historians, even though there is no doubt that their chief livelihood lay elsewhere.

But, interestingly, when scholars wrote specifically as architectural historians they were sometimes criticised as being 'detached', in other words not making clear the implications of their research for the practicalities of everyday life. The *Architectural Review*, for example, in reviewing his book on John Nash, scolded John Summerson – otherwise very active indeed in the debate about new architecture and conservation – for his 'deliberate understatements', the reviewer's concern (to quote Watkin) being 'to give practical application to architectural history'.[24] Even in the 1960s, the

SAHGB, which by then was the premier group encompassing architectural historians in all the sectors (amateur and professional), felt it inappropriate to opine publicly on the demolition of the Euston Arch, one of the seminal moments in the rise of the conservation lobby – conservation being considered beyond its remit. While individual members of the society were deeply involved in conservation issues, and have remained so, often through the more overtly conservationist bodies, it was only in 2005 that the society formally addressed its relationship (along with that of architectural history) to the protection of the historic environment.[25]

Architects and architectural writers and historians formed the core of the amenity societies, such as the Georgian Group and the Victorian Society, that emerged to combat the growing threats to buildings of those periods.[26] Significant works of architectural history were published and helped to mobilise campaigns against demolition. But there is little direct evidence that these works had much impact on wider public attitudes except where they formed part of the new thinking about history and the built environment that did begin to influence perceptions of neighbourliness and the significance of place, and the role of historic buildings in constructing people's view of the world. Let us take this part of the discussion in two bites: to look first at the impact of this new thinking on our theme, and secondly at the extent to which it influenced conservation policy.

The work of Jane Jacobs on American cities, and especially the old-established mixed community of North End, Boston, Massachusetts, and that of Willmott and Young on the East End of London are two of the best-known catalysts that led to more sympathetic approaches to the historic environment although neither was concerned primarily with conservation per se. These pioneer works, from which flowed an entire genre, celebrated the diversity and workability of ordinary places, and led in time to a re-evaluation of the ways in which people interact with familiar surroundings.[27] Jacobs recognised that the key to successful urban renewal lay not in monolithic clearance and redevelopment but in the careful rehabilitation of neighbourhoods that might be overcrowded and run down but were vital, diverse and worked. Willmott and Young analysed the networks of kinship and neighbourliness in Bethnal Green, which they believed created a close-knit and sustainable society that lived out its life in old and sometimes worn-out buildings and streets; once gone, neither streets nor communities could be reconstructed easily, if at all.

Such thinking cross-fertilised with that of social historians who were increasingly concerned with the aspirations and culture, as well as the living conditions, of local communities.[28] Many historians were recognising that people attached multiple significances to events, neighbours, family and places, and that there was a validity to be derived from these that vied, and perhaps competed, with traditional historical assumptions. It was asserted that every individual has memories and organises their own personal histories, operates in a 'temporal *mélange*' (the words are Lowenthal's) and inhabits a world where 'All human beings are practising historians'.[29] Such shifts in scholarly perspectives have been called the 'new scholarship on memory' and involve, among other things, the study of conservation itself as a form of cultural practice. This challenges the old art- (and architectural-) history orthodoxy that distinguished (in terms of both definition and the attribution of significance) between art and simple things, an orthodoxy upon which the whole edifice of conservation has traditionally relied.[30]

Intellectual trends such as these impacted on official thinking. The 1960s saw a growing concern on the part of government about the effects of development, especially on historic towns. Recommendations for action contained in key documents such as *Traffic in Towns* and the conservation assessments of Bath, Chester, Chichester and York were stimulated directly by enlightened criticism in the architectural and planning press and pushed forward by committed politicians.[31] This change of heart was not simply a phenomenon that ran parallel to the intellectual developments outlined above. Such a change could not have taken place without them. The Civic Amenities Act 1967, which enabled local authorities to set up conservation areas, and the Town and Country Planning Acts of 1968 and 1971, which greatly strengthened the hand of local authorities in the field of conservation, created an official climate that was far more receptive to the needs of the historic environment and grass-roots opinion. By the early 1970s, the mass clearances of terraced housing had largely ceased. The scope of what constituted a significant historic building expanded to include industrial monuments – the National Trust allowed industrial monuments to be added to its portfolio for the first time in 1963 – and these along with vernacular buildings were gradually added to the statutory lists. All this reflected specialists' views of importance, spin-offs from their ever wider-ranging studies, together with a growing public awareness that the historic environment was more than great monuments and incorporated 'the cherished local scene', places that touched their lives more closely.[32] Although the nationally organised listing assessments never formally adopted the American National Register threshold for inclusion – 'exceptionally important within the local context' – in practice they did so, and the number of listed buildings more than doubled during the listing surveys of the 1980s.[33]

Government planning guidance for the historic environment published in the 1980s and 1990s reflected much of the spirit of these changes in attitude. The intellectual and social forces that had unleashed dramatic and effective direct action such as the campaigns to save Covent Garden and Spitalfields permeated deeply into the official psyche, too. Gently expressed, the import of government policy was profound:

> The physical survivals of our past are to be valued for their own sake, as a central part of our cultural heritage and our sense of national identity. They are an irreplaceable record which contributes, through formal education and in many other ways, to our understanding of both the present and the past.

This much echoed international heritage conventions. What followed went further: the presence of physical survivals

> adds to the quality of our lives, by enhancing the familiar and cherished local scene and sustaining the sense of local distinctiveness which is so important an

aspect of the character and appearance of our towns, villages and countryside.[34]

PPG15 (from which this quotation is taken), as has been said, formed something of a high point for conservation in England in that it stressed the government's commitment to its stewardship of the historic environment; adopted a broad definition of what constituted that environment; set environmental management squarely in the context of sustainable development; and emphasised the importance of systematic expert historic research. It was in PPG15 that the policy of carrying out thematic research on poorly understood building types or periods was established when identifying buildings for listing.[35] This represented a significant break with previous practice, which was based on geographic parish-by-parish surveys. The first thematic listing programme focused on textile mills, but the one that captured the most public interest was the post-war listing programme. This programme throws up a number of relevant issues and it is to these that we now turn.

Conservation and the use and abuse of architectural history

Two sets of pressures upon government ministers persuaded them to adopt a more systematic research-based approach to the protection of historic buildings. The publication of PPG15 and the increase in the number of listed buildings (which had increased fourfold since 1970 and stood at around half a million individual buildings in 1994) alarmed some landowners and those involved in the urban development and construction sectors. Of the many factors they considered unsatisfactory, one is of particular relevance to this discussion: the lack of transparency regarding the selection criteria for the listing of buildings. They had a point. There was no formal communication with owners of buildings being considered for listing until they received formal notification that the deed was done. PPG15 stated that there was no intention of consulting with owners prior to listing.[36] As angry owners made representations to ministers so, too, did architectural historians. Worried by the under-representation of modern buildings on the lists and, consequently, their vulnerability, a number of historians and architects canvassed the government to extend the range of eligibility for listing to the post-war period. The government's first response, in 1988, had been to open up the field by way of an informal public competition through the pages of the press, but the results of this created more, rather than less, dismay among experts since only a small selection of the winning entries were listed and, furthermore, on the basis of very opaque selection criteria. Pressure mounted for a more systematic and transparent system of selection of modern buildings and the post-war listing programme was launched in 1992.[37]

The post-war listing programme was constructed around the primacy of research. As noted above, the driving principle was to base recommendations for listing on systematic research projects focused on building types (such as schools and public housing but also including planning categories such as New Towns). A panel of experts, chaired by two respected architectural historians, first Ron Brunskill and later Bridget Cherry, oversaw the programme. Both English Heritage historians and external experts carried out research that from an early stage was put out to public consultation. Research was designed to 'identify key exemplars for each of a range of building types . . . and to treat these exemplars as broadly defining a standard against which to judge proposals for further additions to the list'.[38]

Because the key role of architectural history was emphasised in all the official documents, it soon became the principal target of criticism. First the function of 'exemplars' was misunderstood. Some claimed that the research was definitive rather than indicative in the sense that the recommendations for listing represented everything of significance within the type rather than providing benchmarks, which was their primary purpose. Historians would never have claimed that level of omniscience but the misconception of a simple word bred scepticism about the rigour of the research especially when new recommendations for listing were proposed later.

Related to rigour was the issue of objectivity. The difficulties of assessing the significance of things or events of the recent past had long been recognised. Technically, ministers could list buildings that had just been completed if they considered that they met the selection criteria but in practice they imposed a self-denying ordinance of thirty years. One of the publications that launched the programme, *A Change of Heart*, justified the rule succinctly: 'The thirty-year rule is not without difficulties, but it has the merit of enforcing perspective. The ups and downs of architectural reputations since 1945 warn us that we would be unwise to accept all that is currently fashionable in architecture as of permanent value.'[39] Exceptionally, buildings were listed that were less than thirty years old and in such cases the safeguards were strengthened: they had to be deemed outstanding and under immediate threat; English Heritage's commissioners were required to view the case, as did the Secretary of State; and as with all post-war candidates, public opinion was sought. But the involvement of architectural history in the service of so publicly exposed an initiative as the post-war programme, operating in an area some saw as belonging more to fashion than to academe, threatened to demean it in the official eye. In 2001, the government, for the first time, turned to another body to advise on listing. From that point the Commission on Architecture and the Built Environment (CABE) enjoyed a formal role (alongside English Heritage) in the assessment of 'special historic or architectural interest' – the criteria for listing.[40]

The tension between perceptions of fashion (deemed transitory) and of architectural interest (cool judgements about enduring cultural value) came to a head in a related issue: fitness for purpose. One way both to challenge and exploit the value of architectural history was to contest the listing of a building on the grounds that it did not work. This goes to the heart of architecture, which is about accommodating function in an aesthetically coherent way. Historians might insist that a building had historic value even though it was no longer fit for purpose. The nub of the issue was less whether a building could cope with new uses than whether it ever fulfilled its original brief effectively.

The debate about fitness for purpose intensified when Pimlico School, Westminster, was considered for listing. This extraordinary building of 1966–70 was one about which few people who saw it had no opinion. Pevsner considered it 'wild and weird . . . restless, aggressive, exciting' – a view maintained by those who revised his *Buildings of England* volume in 2003.[41] But all agreed that it had major design failures which created a greenhouse effect, making parts of the building unusable for some of the year. Pevsner, furthermore, had asserted that it was a modish design that would 'date quickly'. Architectural history was mobilised to argue both for the intrinsic interest of the building and for its inherent failings. After extensive debate, English Heritage recommended in 1996 that the building was not listable in a high grade (which it had to be under the 'thirty-year rule') and might not be of Grade II quality. In 2002, it recommended listing in Grade II.[42] Such apparent inconsistency in the treatment of a single case threatened to bring the process of assessment into disrepute and with it the use made of architectural history.

The post-war listing programme also came under attack from within the architectural history establishment itself. In *A Change of Heart*, Andrew Saint set out the range of architecture produced in the post-war years – 'of unsuspected, often bewildering variety, richness and inventiveness' – and identified three strands: 'the modernism of method, the modernism of style and the modernism of good manners'. The last of these small-'m' modernisms was characterised by its 'lack of dogma and rhetoric, its refusal to jettison tradition'.[43] Such a catholic approach to the architecture of the period provoked those who still held a deep attachment to Modernism and felt that widening the listing net to include what Summerson had once called middle-brow and cosy was to debase the currency. In a hard-hitting attack on *A Change of Heart* and the post-war listing programme, Nigel Whiteley, a university-based art historian, interpreted Saint's use of the term modernism as misappropriation in the service of a hidden and conservative agenda. Modernism was being redefined 'so that it fits the sort of *identity* of Englishness which heritage organizations frequently promote. The quirky, the idiosyncratic, the eccentric and the finicky are taken as significant and characterful . . . it is misleading and bad history.'[44]

Attending to everyone's past

This discussion about English Heritage's post-war listing programme has helped bring into focus a number of issues that became the subject of heated debate in the later 1990s: openness by means of publicising the selection criteria and justifying judgements about significance; the primacy of research; the need for rigour and objectivity and exposing these to challenge; distinguishing between passing fashion and perennial values; and the uses to which research is put.[45] The programme helped raise public awareness of the qualities of post-war architecture through the media, exhibitions and publication. By 2000, a MORI poll found that 75% of people believed the best of our post-war buildings should be preserved (rising to 95% of the sixteen to twenty-four age group).[46] But by the same token, this heightened scrutiny and debate served to muddy the waters between architectural history and the protection of the historic environment on the one hand and architectural history and the promotion of heritage on the other. Within the wider context of what might be called the democratisation of heritage alluded to above, architectural history in the service of conservation was now confronted with a number of challenges. We turn to these in the final section of this chapter.

In developing a policy towards the historic environment at the dawn of the twenty-first century, the government has had to strike a balance between maintaining a credible system of protection based on sound and defensible criteria; broadening appreciation of that environment through outreach and education; and acknowledging all the diverse and sometimes competing values that people attach to the historic environment. It is a minefield. Underlying some critiques is the fear that heritage will come to reflect a stodgy, fuddy-duddy establishment view that avoids thorny issues and self-questioning or, at the other extreme, to represent the views of a small academic elite that carry no resonance with the majority of people. Others fear that popularisation will invoke the deadening hand of commoditisation and package 'the environment as entertainment, or nostalgia, or never-never-land'.[47] There is growing recognition, reflected in the criteria public bodies such as the Heritage Lottery Fund now use to fund heritage programmes, that a multitude of values need to be taken into account before measuring public benefit. Historians, in order to bring some order to the study of how people perceive their pasts, have categorised their endeavours (in so far as they relate to the historic environment) into political or official narratives, popular history and the history of place.[48] Mediating successfully between these categories, it is suggested, may be the key to developing a coherent approach to understanding and managing the physical attributes of what is increasingly called the 'cultural resource'.

It was into this minefield that English Heritage strode purposefully with the biggest consultation exercise ever held on the meaning and future of the heritage. The results of this exercise (which represented the views of the heritage sector and involved around 180 experts and over 600 organisations) resulted in the publication in 2000 of *Power of Place*.[49] The document accepted the importance of research (although the section entitled 'The first precondition: knowledge' came at the very end of the document) but its main thrust was towards inclusivity. Reflecting both the new historical thinking and the priorities of government, it asserted that people valued the historic environment for its 'meanings, its beauty, its depth and diversity, its familiarity, its memories', and declared that one of the challenges was that 'many [people] feel powerless and excluded'. It continued: 'The historic contribution to their group in society is not celebrated. Their personal heritage does not appear to be taken into account by those who take decisions' and urged that 'the heritage sector find out what people value about their historic environment and why, and take this into account in assessing significance'.[50] The primacy of expert opinion was toppled.

The government's response to this major consultation exercise, *The Historic Environment: A force for our future*, similarly recognised the diversity of perceptions about the historic environment and saw the latter as 'something that can bring communities together in a shared sense of belonging'. The document laid out government policy (with fifty-four action points) and paid particular attention to the role of good-quality research, which it saw as 'vital to inform the direction of policy'. English Heritage was tasked to bring some order to what was seen to be a disorganised research sector: 'The Government believes that a coordinated approach to research is essential if its full benefit is to be realised.'[51] Precisely how (and indeed whether) this will work it is, at the time of writing, too early to say. English Heritage's research strategy was published in 2005. This systematised its research activities (and, as the single biggest national funding body for research on the historic environment, that of the sector) into programmes that related closely to the government's principal policy objectives. It showed that of English Heritage's research budget (around £7.5 million per annum), the bulk (88%, about £6.6 million) was spent on 'discovering, studying and defining historic assets and their significance' – one of the core activities of the body, conventionally defined. As to the study of 'other values' attached to the historic environment, this was subsumed within a category that also included research on economic benefits and accounted for 4% (£262 000) of the whole.[52] In other words, at present, a very small proportion of the research budget goes on understanding alternative or wider community attitudes to the historic environment and heritage value. Yet more and more, it appears, this is what counts.

We have moved a long way from the first Victorian architectural historians and the early days of the SPAB when large swathes of what we would now regard as conventional heritage such as historic private houses were considered to be off limits to conservationists. The expansion of the statutory lists of historic buildings, identified for protection by architectural historians, has generally met with public acceptance. The Heritage Protection Reform, one of the fruits of *A force for our future* (and covered elsewhere in this volume), by developing the criteria for selection and putting them out to consultation, is further exposing the premises of architectural historians to public scrutiny.[53] As the more holistic approach to the management of the historic environment, promulgated in these policy documents, takes root, historians and conservationists will continue to have their heels bitten by those who see them as 'the heritage brigade', hell-bent on protecting everything.

More difficult to challenge or assimilate, however, is policy-makers' espousal of a view of heritage that is so widely defined as to be nebulous. Heritage ministers made clear the extent to which the goal posts have moved in two statements. At a Royal Geographical Society conference on 'capturing the value of heritage' in January 2006, David Lammy (the Minister for Culture) said: 'It is up to people, institutions and civil society to determine their own conceptions of heritage.' On the same occasion, Tessa

Jowell, the Secretary of State, declared: 'Instead of experts making all the decisions, experts would share their knowledge with the public, and facilitate people making more of their own informed judgements.'[54] In such a world as this, architectural historians will have to more actively engage outside the 'official narratives' and in the arenas of popular history and the history of place if they are to play a prominent role in the conservation of the historic environment.

Endnotes

1   See the debate between Tim Williams et al., in 'Closing the divide: a discussion about archaeology and conservation', in The Getty Conservation Institute Newsletter **18**, 1 (2003), 11–17.

2   For a recent account of the effects of changing attitudes to monuments, see Anna Keay, 'The presentation of guardianship sites', Transactions of the Ancient Monuments Society 48 (2004). The interpretative problems associated with discrete intervention were raised by David Brock at a joint Society of Architectural Historians of Great Britain/Institute of Historic Buildings Conservation conference held in March 2005: see note 25 below.

3   Malcolm Airs, 'Architectural history and conservation education', unpublished paper given at the March 2005 SAHGB/IHBC conference: see notes 2 above and 25 below. I am grateful to Professor Airs for making a copy of his paper available to me.

4   Department for Culture, Media and Sport, The Historic Environment: A force for our future (DCMS, London, 2001), p. 15 (1.12); for the AHRC Landscape and Environment Programme, see www.ahrc.ac.uk; English Heritage, English Heritage Research Agenda: An introduction to English Heritage's research themes and programmes and Discovering the Past, Shaping the Future: Research strategy 2005–2010 (both English Heritage, London, 2005).

5   Avista Forum Journal **13**, 2 (2003), 27–8: Report on Kalamazoo conference panel discussion on interdisciplinary studies.

6   C.R.J. Currie, 'The unfulfilled potential of the documentary sources', Vernacular Architecture 35 (2004), 1–11.

7   Frank Salmon, 'The society and the discipline: where next?', Society of Architectural Historians of Great Britain (SAHGB) Newsletter 81 (Winter, 2003–4), 1–2.

8   Richard Tomkins, 'Why technology changes everything, even my trousers', Financial Times, 3 January 2006.

9   David Stocker, 'The pre-history of conservation', in Reconstruction and Regeneration: The philosophy of conservation, Proceedings of the IHBC Canterbury School (1997), pp. 6–12.

10 David Watkin, The Rise of Architectural History (Architectural Press, London, 1980), pp. 71–3. Freeman is also given full coverage in what is still the best single account of the architectural writing of this period: Nikolaus Pevsner, Some Architectural Writers of the Nineteenth Century (Clarendon Press, Oxford, 1972), especially pp. 101–2.

11 Edward A. Freeman, A History of Architecture (London, 1849), pp. xi–xii, xix, 11.

12 George Aycliffe Poole, A History of Ecclesiastical Architecture in England (London, 1848), especially pp. 341 passim.

13 Ibid., p. viii.

14 James Fergusson, History of the Modern Styles of Architecture (Murray, London, 1862). This was extensively enlarged by Robert Kerr (3rd edn, 2 vols, 1891), bringing the story bang up to date.

15 The quotation and what immediately follows are taken from Colin Cunningham, 'Practicality versus conservation: Alfred Waterhouse and the Cambridge colleges', *Architectural History* 37 (1994), 131–52.

16 Chris Miele (ed.), *From William Morris: Building conservation and the Arts and Crafts cult of authenticity, 1877–1939*, Paul Mellon/Yale Studies in British Art 14 (Paul Mellon Centre, New Haven and London, 2005), p. xi.

17 A Watch Committee was set up to identify historic buildings under threat in Greater London and agitate for their preservation; see Alan Crawford, *C. R. Ashbee: Architect, Designer & Romantic Socialist* (Yale University Press, New Haven and London, 1985), pp. 57–66. The terms of the Royal Commission charter appear in all the published volumes of that body, later termed the Royal Commission on the Historic Monuments of England (RCHME).

18 Thomas Garner and Arthur Stratton, *The Domestic Architecture of England during the Tudor Period* (B.T. Batsford, London, 2 vols, 2nd edn, 1929), p. vi. (The quotation appeared in the first edition of 1910 and was reprinted in the second.)

19 See e.g. Michael Bartholomew, *In Search of H. V. Morton* (Methuen, London, 2004); other writers are covered in the *Dictionary of National Biography*, e.g. John Massingham (by Malcolm Chase), Stuart Mais (by Bernard Smith) and Ruth Manning-Sanders (by Veronica Hurst).

20 Miele, *From William Morris*, pp. 4–6.

21 First appearing in the *Architectural Review* in 1955, it was published separately as Ian Nairn, *Outrage: Counter-attack against Subtopia* (Architectural Press, London 1957); see Elain Harwood and Alan Powers (eds.), 'The heroic period of conservation', *Twentieth Century Architecture* 7 (2004).

22 Watkin, *The Rise of Architectural History*, pp. 144–5.

23 John H.G. Archer, 'Red and black: the origins of the Society', *SAHGB Newsletter* 80, (Autumn 2003), 1.

24 Watkin, *The Rise of Architectural History*, p. 131.

25 Archer, 'Red and black', p. 4; Martin Cherry and Emily Gee, 'SAHGB/Institute of Historic Building Conservation (IHBC) Symposium', *SAHGB Newsletter* 87 (Spring 2006), 9–10.

26 Neil Burton, 'A cuckoo in the nest: the emergence of the Georgian Group', in Miele, *From William Morris*, pp. 237–58; Gavin Stamp, 'The art of keeping one step ahead: conservation societies in the twentieth century', in Michael Hunter (ed.), *Preserving the Past: The rise of heritage in modern Britain* (Sutton Publishing, Stroud, 1996), pp. 76–98.

27 Jane Jacobs, *The Death and Life of Great American Cities* (Vintage, New York, 1961); Michael Young and Peter Willmott, *Family and Kinship in East London* (Penguin, London, 1972).

28 The association of historic neighbourhoods with social cohesion as a dynamic in conservation is still alive, for instance, in responses to the government's Housing Market Renewal or Pathfinder programmes; see Stefan Muthesius, 'The Victorian terrace: an endangered species again?', *The Victorian* 21(March 2006), 4–8.

29 David Lowenthal, *The Past is a Foreign Country* (Cambridge University Press, Cambridge, 1985), p. 186; Gerda Lerner, *Why History Matters: Life and Thought* (Oxford University Press, New York, 1997), p. 199.

30 David Glassberg, 'Presenting memory to the public: the study of memory and the uses of the past', *Cultural Resource Management* 21, 11 (1998), 7. The classic statement on the institutional ownership of judgements such as these remains Pierre Bourdieu, *The Fields of Cultural Perception* (Polity Press, Cambridge, 1993), particularly chapter 10: 'The historical genesis of a pure aesthetic', pp. 254–66. Key studies in this vein include Lowenthal , *The Past is a Foreign Country*; Raphael Samuel, *Island Stories*, 2 vols, *Unravelling Britain* and *Theatres of Memory* (Verso, London, 1994, 1998); Peter Mandler, *The Fall and Rise of the Stately Home* (Yale

University Press, New Haven and London, 1997); Miele, *From William Morris*, is particularly interesting on this, particularly the preface and chapter 1, 'Conservation and the enemies of progress', pp. 1–29.

31 The best recent coverage of these developments is Hunter, *Preserving the Past* (see note 26 above), and John Delafons, *Politics and Preservation: A policy history of the built environment 1882–1996* (Routledge, London, 1997). Wayland Kennet, *Preservation* (Temple Smith, London, 1972), remains an excellent account from the perspective of a political activist of the time.

32 The phrase appeared first in Department of the Environment Circular 8/87 and remained in PPG15.

33 The National Register criteria are discussed interestingly in the context of urban renewal by Richard Longstreth, 'The difficult legacy of urban renewal', *CRM: The Journal of Heritage Stewardship* **3**, 1 (Winter 2006), 6–23.

34 PPG15, p. 1 (1.1).

35 PPG15, p. 27 (6.12).

36 PPG15, p. 26 (6.5). English Heritage was expected to liaise with the local authority concerned, however, in the case of thematic list reviews.

37 The post-war listing programme and some of the associated politics are covered in Martin Cherry, 'Listing twentieth-century buildings: the present situation', in Susan MacDonald (ed.), *Modern Matters: Principles and practice on conserving modern architecture* (Donhead, Shaftesbury, 1996), pp. 5–14. See also Diane Chablo, 'The listing of post-1939 buildings', *Conservation Bulletin* 4 (February, 1988), 5–6; Diane Kay, 'Post-war listing: an update', *Conservation Bulletin* 20 (July, 1993), 12–13.

38 PPG15, p. 27 (6.12).

39 Andrew Saint, *A Change of Heart: English architecture since the war – a policy for protection* (Royal Commission on the Historic Monuments of England, London, 1992), p. 27.

40 *A force for our future*, p. 35 (4.8).

41 Nikolaus Pevsner, *Buildings of England: London 1*, 3rd rev. (Yale University Press, London, 1973), pp. 114, 515; Simon Bradley and Nikolaus Pevsner, *Buildings of England: London 6* (Yale University Press, London, 2003), pp. 764, 769–70.

42 There was considerable media cover of this case in the autumn of 1996. The details are laid out in Mr Justice Gibbs's judgement, *John Bancroft v. Secretary of State for Culture, Media and Sport, Westminster City Council*, 15 June, 2004. Ministers issued a certificate of immunity against listing in 2003.

43 Saint, *A Change of Heart*, pp. 5, 13. Saint scrupulously uses a lower-case 'm' for modernism throughout.

44 Nigel Whiteley, 'Modern architecture, heritage and Englishness', *Architectural History* 38 (1995), 235. Whiteley consistently uses an upper-case 'M' for Modernism throughout (except when quoting Saint).

45 'Private citizens are of course entitled to save their own past, but when preservation becomes a public act, supported with public funds, it must attend to everyone's past.' Herbert J. Gans, quoted in Dolores Hayden, *The Power of Place: Urban landscapes as public history* (MIT Press, Cambridge, MA, 1999), p. 4.

46 English Heritage, *Power of Place: The future of the historic environment* (English Heritage, London, 2000). The results of the MORI poll ('Attitudes towards the Heritage') are summarised on p. 4.

47 Ada Louise Huxtable, *The Unreal America: Architecture and illusion* (New Press, New York, 1997), p. 2. The literature on this theme is now huge; the current generation of such works in Britain started with Robert Hewison, *The Heritage Industry* (Methuen, London, 1987), and Patrick Wright, *On Living in an Old Country* (Verso, London, 1985).

48 Glassberg, 'Presenting memory to the public', 8.

49 *Power of Place*, Chairman's Foreword, p. 1.

50 *Ibid.*, recommendation 9, p. 27.
51 *A force for our future*, p. 15 (1.12).
52 *English Heritage Research Agenda*, pp. 4–5. For full publications details, see note 3 above.
53 Department for Culture, Media and Sport/Office of the Deputy Prime Minister, *Revisions to Principles of Selection for Listing Buildings: Planning Policy Guidance Note 15 – consultation document* (DCMS, London, 2005).
54 www.culture.gov.uk/global/press-notices/archives-2006

3 Conservation and authenticity

Martin Robertson

Listed building legislation has been active since the Town and Country Planning Act 1947, but in practice buildings have only been protected from change in any thoroughgoing way since the Town and Country Planning Act 1968 made it necessary to apply for listed building consent before undertaking any works. This Act neatly coincided with the beginning of the enormous expansion in the numbers of listed buildings that resulted from the resurvey of England 1966–91, while Scotland, Wales and Northern Ireland also began the long-drawn-out process at about this time.

The number of listed buildings in England during this period rose from about 100000 to about 500000 by the mid-1990s, and the increase was mirrored in relative terms in the other countries – 500000 for an English population of about 50 million; 30000 for a Welsh population of about 3 million – suggesting a ratio of 1:100 listed structures to people. This naturally placed a huge additional burden on local planning officers, very few of whom had enjoyed any kind of conservation training and who were likely to know very little about the considerable numbers of specialist buildings and structures now in their care. The resurvey itself trained many people in the necessary knowledge about the buildings – up to 200 people worked on it in England alone – and these became available for employment in the areas they had come to know well as the resurvey in their locality came to an end. Many are still in post today. In addition university postgraduate courses in building conservation began to appear (at the Architectural Association, Bath, Oxford Brookes, York and elsewhere) offering a formal qualification in the subject. Before these commenced there had only been academic courses in architectural and vernacular building history (Manchester, Reading, York) or in art history with some architectural history (Edinburgh, London, Oxford), and while these had provided many of the most knowledgeable of the fieldworkers for the resurvey, looking after the buildings so identified in the future would make very different demands.

It must be remembered that listing and thus the commitment to the protection of the British built heritage was, and continues to be, part of a due democratic process enshrined in all the proper panoply of parliamentary law-making. A working life with listed buildings has shown very clearly that a large proportion of the population supports the ideas behind listed buildings and conservation, as long as the buildings belong to somebody else. It is when the legislation appears to threaten property owners' personal liberty that the system becomes an 'iniquitous scandal', and so on.

And yet, of course, there has been much justification in the complaints of bureaucracy, contrariness and apparently downright stupidity that are common in tales of attempts to progress perfectly reasonable alterations to listed buildings. The state, and particularly the local planning authority, is seen as putting an unfair restriction on the owners of listed buildings who are thus unable to use their buildings as they wish and to make the changes they consider they need, and who know very well that what they want to do has been allowed in the past in other areas, in the next street, or even next door. There are, of course, usually good reasons for this having happened, but the reasons are often ones that the unfortunate owner does not wish to understand and it becomes an 'us against them' situation.

So this brings us to the question of what it is we are trying to do by listing buildings and other structures as being of special architectural or historic interest, and also by scheduling them as Ancient Monuments. In fact, the principal reason and the aim of the whole process is clear: to preserve and enhance the nation's heritage of fine architecture and other important structures for the benefit of all, both now and in the future. The question of how that aim is to be achieved is an altogether more difficult one.

The official government advice on the care of historic buildings is Planning Policy Guidance Note 15: *Planning and the Historic Environment* (PPG15), published in September 1994. This has general policy in Part 1 and in Part 2:7, 'The upkeep and repair of historic buildings', and there is also Appendix C: 'Guidance on alterations to listed buildings'. This PPG was the response of the Department of National Heritage to its new position as the caretaker of Britain's historic buildings and monuments, having taken over from the Department of the Environment in 1992. It replaced the previous DoE circulars 23/77 and 8/87, which had carried much less detailed guidance. Its mission statement is quite plain:

> It is fundamental to the Government's policies for environmental stewardship that there should be effective protection for all aspects of the historic environment. The physical survivals of our past are to be valued and protected for their own sake, as a central part of our cultural heritage and our sense of national identity. They are an irreplaceable record which contributes, through formal education and in many other ways, to our understanding of both the present and the past. Their presence adds to the quality of our lives, by enhancing the familiar and cherished local scene and sustaining the sense of local distinctiveness which is so important an aspect of the character and appearance of our towns, villages and countryside.[1]

> The Government has committed itself to the concept of sustainable development – of not sacrificing what future generations will value for the sake of short-term and often illusory gains.[2]

This message is clear, but it is hardly a new one. The key part of the above to which we all respond is:

> The physical survivals of our past are to be valued and protected for their own sake, as a central part of our cultural heritage and our sense of national identity. They are an irreplaceable record.

It was much the same in 1877 when William Morris wrote his *SPAB Manifesto*:

> If . . . it be asked us to specify what kind of amount of art, style or other interest in a building, makes it worth protecting, we answer, anything which can be looked on as artistic, picturesque, historical, antique or substantial.

Historic buildings must be looked after with honesty of purpose and sensitivity to their true character:

> Thus, and thus only can we protect our ancient buildings, and hand them down instructive and venerable to those that come after us.

And, after forty years of government advice, it is much the same today; for example, wise words and uncertainty of purpose from Loyd Grossman, a former English Heritage Commissioner:

> We can legitimately regard England as the birthplace of the modern conservation movement, and as a result there is a large amount of academic and practical knowledge about conservation and a strong cultural bias towards its value. Equally it is important to recognize that England has been, and to a certain extent remains a strongly hierarchical society as well as an increasingly cosmopolitan one, with many competing ideas about what heritage is and indeed to whom it belongs.
>
> . . . There is of course the constant pressure to somehow value the contribution that heritage makes to society, and while it is possible to quantify the role heritage plays in economic regeneration (the most obvious examples can be seen in the revival of our great regional cities), it is difficult, if not downright impossible, to say what the exact worth of heritage is in terms of building citizenship, spiritual values or a sense of meaning and belonging.[3]

The surprising point is that Loyd Grossman still has to say it. William Morris was a lone voice calling out against the god of property in the very different society of Victorian England and telling us heritage is important and that it belongs to everyone. This has become a world truism and yet, despite this, it is still under attack from all sides. Every year there are more protected buildings and structures than ever before, but despite investment in them via education, management, grants, amenity societies and the huge number of visits to historic towns, country houses, industrial remains and historic landscapes, there are still too many examples of neglect, abuse and misuse to be seen in every part of the country. The conservation of historic buildings and structures through the official channels of listing and listed building consent does not work as well as it should; a root cause of this is the mixed messages used in the promotion of conservation, and these messages often concern the philosophy of why we do it in the first place.

It is obvious that the older a building is the more likely it is to have suffered significant change. Much change, however, is undocumented or its provenance is forgotten, and many of the buildings protected today have lost a degree of authenticity through change and restoration that now makes them of less historic value than before. Almost every major house will have had at least one fire – Prior Park in Bath, for instance, has had two, in 1836 and in 1991 – and when you think of how they were heated and lighted through most of their lives it is hardly surprising. When they are rebuilt they are restored, or updated, or altered, or added to, or all of these, and at the time of the next fire it becomes a puzzle as to what to do, what to return to. Do you restore what was there before the fire or do you return to what the first designs were, or may have been? Thus Prior Park in 1991–5 became an amalgam of 1735, adaptation and change post-1836, and repairs, modernisations and reinstatement of work of both 1735 and 1836. So this is now a house of three periods, which has been in its life a country house, a seminary and a boarding school, as well as twice being a ruin. Although this was the finest Bath house of its time and ostensibly still is, it is also a building that, in its current complete state, has existed only for ten years and would, in parts, be unrecognisable to all its previous owners. This is the essential phoniness of historic buildings. Yes, their development and the changes they have undergone over time are all part of their history and interest, and should be respected; but all too often an arbitrary/informed decision is taken by those in charge of the heritage that a particular appearance is the right one.

To take a very obvious example, Britain's most famous monuments, Stonehenge and Avebury, are recognised for their historic importance and cultural value as World Heritage Sites. This is perfectly correct; they are wonderfully evocative and interesting monuments. But genuine and uncomplicated relics of the past they are not. Both monuments were reconstructed more than once in the pre-Roman period and then spent centuries in quiet decay with their stones being robbed for other purposes, until antiquarians first began to notice and record them in the seventeenth century. Inigo Jones attempted the first drawn reconstruction of Stonehenge with the surprising conclusion that it had been square in form, but John Aubrey and William Stukeley began to add accuracy to surmise. Stonehenge is, luckily, some distance from any village so stone robbery was not as important a problem as at Avebury, but the best-known depictions, made by Constable and Turner in the early nineteenth century, show the monument as they saw it, not as we do. In 1957 the Ministry of Works chose to re-erect a number of the stones to return the monument to the appearance it had in the eighteenth century, and the alteration is immediately apparent if you compare today's appearance with the Constable watercolours. The question might well be asked: 'Why is its eighteenth-century appearance the correct one?'

It is the same with Avebury. Samuel Pepys visited in 1668 and saw 'a place trenched in . . . with great stones pitched in it some bigger than those at Stonage [Stonehenge] to my great admiration'. But we do not see the stones that Pepys saw. In 1724 William Stukeley recorded that the surviving stones were being smashed and taken to be built into the village houses, where they still are. The stones we see today are mostly those ones that had been buried in the Middle Ages on the orders of the Church and were found and re-erected by Alexander Keiller in the 1930s. Our Avebury, in its current appearance, has thus only existed since 1939 and at no other period in its history. This paradox was further illustrated in 2005 by the discovery that one of the Keiller stones had been erected upside down, so the story may well have yet another twist to come.

A far more wide-ranging example is the problem presented by the effect that changes in standards of comfort have had on the listed buildings of Bath, though this can be repeated anywhere with terraces, crescents and other multiple housing groups. Most of the eighteenth-century houses of Bath were built speculatively on upwards of ninety-year leases, and very often, as the leases ran their course, the houses became progressively more neglected. Thus, when they were sold freehold at the end of the nineteenth century, they were both old-fashioned and in poor repair. The addition of plumbing to the rear and front facades and the replacement of the windows with plate glass sashes was wholesale. When listed in the 1950s almost all these buildings had been mutilated in this way and the conservation movement has since found it very hard to adapt to the problem. The Georgian Group held its annual conference in Bath in May 1948 and was keen to recommend the removal of unsightly additions and the restoration of missing features. It said of plumbing:

> The introduction of baths, wash-hand basins and WCs into Georgian buildings created a problem which in the majority of cases has been very badly solved. Down-pipes of all descriptions were fixed to the rear and front of the buildings with no thought to the architectural composition, and indeed, in some cases actual physical damage to the architectural features has resulted . . . By careful design and forethought it is possible in most cases for pipes of this description to be contained in ducts within the actual building itself.

Of 'cliff-hanger' bathrooms:

> These unsympathetic and ill-conceived excrescences are the direct result of the advance of civilization, or adaptation of single properties for multi-family use. These appendages could have been avoided if the speculator had realised that good design is the cheapest in the end.

And of lost glazing bars:

> Very few properties in Bath now have their original glazing bars. The removal of glazing bars completely alters the whole scale of a building and it is felt that a campaign for the reinstatement and restoration of glazing bars would serve a useful purpose.[4]

Sixty or so years later the plumbing has been improved, the stonework cleaned and features restored in many of the houses. Glazing bars, however, have gone from being actively encouraged by grant and deed to being actively discouraged through refusal of listed building consent. In the space of any central Bath street now there are both glazing bars and plate glass, original sill heights and lowered ones, windows with twelve panes and others with fifteen, glazing bars with mouldings that might be called correct and others that could not – all approaches that were considered appropriate at one time or another. The result of all this, as with Avebury, is to give many of these buildings an appearance today that they have had for only a very short part of their existence.

The realisation that so many monuments and buildings are today a masquerade of reality, the result more of art, nature and history combined, begs the question once again of what we are trying to achieve through their continued preservation. PPG15, as the official government arbiter of taste and history, should be the chief guide. It has worked well since 1994, but not as well as it could have done. It tries to give a proper and useful sense of pragmatic reality, but this is often disregarded in use by those who choose a more rigid and purist interpretation better suited to their own agendas.

The word 'authentic' and the concept of authenticity do not appear in PPG15, but they perhaps should. It is this document, and the advice it contains, that is the principal tool of the local authorities in formulating their conservation policies in their local plan, and it is this document that English Heritage uses to measure the local plan policies against government policy in formulating their advice to the government. The words that are used in PPG15 are 'traditional', 'sympathetic' and 'respect' – safe 'hands-off' words that are not really positive – while the dangerous concept word 'original' is used very occasionally. Perhaps 'authenticity' could provide a better approach that would give the system a clearer idea of what conservation is trying to achieve.

Might authenticity then be a useful concept in connection with the conservation of historic buildings, monuments and structures? What might it mean in this connection? Does it refer to solid virtues of honesty, usefulness, and traditional values and construction techniques? Or does it now refer to contemporary qualities of the twenty-first-century lifestyle as recommended today by so much in the media? 'Authentic' is a word often used by commercial organisations recommending their products; Marks & Spencer and the National Trust for Scotland are just two examples among many, with varying degrees of absurdity, and yet the dictionary demonstrates that authenticity is a concept that has a very real relevance to our perception of the purpose of historic buildings conservation. Authentic means 'honest', surely a self-evident virtue in conservation terms, for what are most historic buildings if not the honest use of locally available materials designed to reflect the social aspirations of their owners? A great deal of advice is now available on how to repair historic buildings and how to convert them to new uses for a modern age, but too little advice on the real purpose of doing this and the values, both practical and emotional, that can be gained from a job well done when a building proves to be in harmony with itself, its surroundings and its owners. Hundreds of thousands of listed buildings have been repaired and converted during the last fifty years, with or without listed building consent. How often, among all these, is the result a truly successful one?

'Authentic' – adjective

Of undisputed origin
  Made or done in the traditional or original way or in a way that faithfully resembles the original
  Based on facts, accurate or reliable

(Continued)

(In existential philosophy) *relating to or denoting an emotionally appropriate, significant, purposive and responsible mode of human life.*
  Origin – late Middle English via Old French from late Latin 'authenticus' from Greek 'authentikos' meaning 'principal' or 'genuine'
  New Oxford English Dictionary, Oxford University Press, 2000

This dictionary definition does suggest that 'authentic' offers a sound basis for the whole philosophy of historic buildings and monuments conservation.

The principles for historic buildings conservation work might be stated thus. Conservation work on historic buildings and monuments should only be undertaken if

1. it is based on accurate and reliable information
2. it uses traditional methods of repair where possible
3. it leads to a historically and emotionally satisfying, honest, appropriate and responsible result.

The first two of these principles are obvious and have been stated many times before but there is as much evidence today as there has ever been to suggest that they are often not followed. It is, however, the third principle that we are mainly concerned with here.

Britain has now achieved one of the most comprehensive systems of heritage protection of any country, but too many listed buildings are still threatened with inappropriate repair and alterations and too many end up in a condition that is not 'historically and emotionally satisfying'. This was highlighted by the publication in 2005 of the Countryside Agency's report on the future of traditional agricultural buildings.[5] This suggested that within twenty years the remaining stock of unaltered traditional farm buildings will come under increasing pressure for conversion to housing and will, in due course, have been either converted or demolished. Therein lies the problem for buildings that have lost their use – agricultural, industrial, institutional, military, whatever. Now, as then, 'We know from bitter experience that the disused becomes the derelict and soon the irretrievably lost.'[6]

The essential dilemmas in this situation were robustly addressed by a columnist in *The Independent* newspaper:

The farmers themselves are too impoverished to pay for upkeep, so quite often they apply for change-of-use planning permission. The result, more often than not, it is claimed, are those horror conversions: shed-houses, or barn-houses designed in a way that is 'fundamentally unsympathetic' and is leading towards the 'suburbanisation of the countryside'.

At first glance, the fate of the nation's barns and cowsheds would seem to rank rather low on a crowded anxiety agenda . . . a tweedy, old-fashioned kind of class snobbery.

. . . Dealing with local planners, I discovered a simple truth. Although the building was not old, nor timber-framed, nor really a barn, it gave the illusion of being all of those things and so, discussing windows, the roof or the chimney, we all had to pretend it was something it was not. The place, I was repeatedly told, must look when it was completed like a converted agricultural building. Anything which resembled a house would have breached the planning regulations.

Here is the source of the Countryside Agency's suburbanisation. It is not over-excited architects working for tasteless and vulgar urbanites who are causing the problems so much as a backward-looking, unimaginative culture that is in thrall to a particular view of the countryside. Hooked pathetically on heritage, the planners insist on houses that are as identical to one another as possible. Each must be a neat, sanitised, environmentally sound imitation of a cowshed, barn or stable.

. . . The horrors are not being perpetrated by people like me but by a drearily unadventurous planning system.[7]

Agreed that this is often the case, but the problem really lies not in the planning system but with the 'drearily unadventurous' way that the rules are interpreted by so many of those entrusted with the protection of our built heritage. Owners are, on the one hand, actively prevented from doing what they want by the local authority conservation officers and, on the other, relentlessly encouraged by the media and the marketing agencies to want the inappropriate. A good example of what the Countryside Agency is talking about was featured in *Bath Life* magazine:

In fact, this lovely barn conversion has already opened its doors to two periodi-cals specialising in traditional homes, inspiring envy in the breasts of thousands of readers across the country – both for its idyllic rural setting and the sensitive renovation process that makes it look as if it has graced the landscape for hun-dreds of years.

In fact when the current owners bought the mid-18th-century hay barn . . . it was all but derelict. They drew up plans that involved dismantling the original building and replacing it with a new barn, albeit one designed on traditional lines, and reusing as much of the original facing stone as possible.

The design managed to satisfy the stringent criteria of the local planning com-mittee, and permission was duly granted. The owners sum up the original vision for their new home as 'barn-like and French', and indeed, the mellow stone building would look just as at home in the lavender-scented fields of Provence as it does in its green and pleasant Somerset setting.[8]

It is evident that this house is an attractive and valuable property that will fulfil its twenty-first-century function very well but, as the article describes, its relationship to the mid-eighteenth-century hay barn from which it sup-posedly sprang is now very tenuous indeed, and it would be sad if, within twenty years as the Countryside Agency claims, all former agricultural build-ings were to look like this. Even more of a contradiction is a house in Wilt-shire where development ideas went full circle and actually provided a new house built to imitate a barn. It must have got planning permission from the local authority and it shows exactly the lack of courage that Terence Blacker complains of. As William Morris said in his *SPAB Manifesto* of 1877: 'A feeble and lifeless forgery is the final result of all the wasted labour.'

Where can conservation go if authenticity is to be the watchword? How best is Britain's irreplaceable stock of historic buildings and structures to be kept in repair and use and to be presented to the future? A Taiwanese dissertation student said charmingly 'Preservation is not about building a better yesterday', but in one way she was wrong. We deal with yesterday's buildings but we do not want yesterday's discomfort and lack of hygiene; in fact, in some ways, a better yesterday is what we do want. We may want the elegance, space and light of a Georgian house; we don't want the draughts or the lack of plumbing and electricity. This is where the compromises come, preserving the character but making the building usable in a reasonably comfortable way. You cannot have a fully authentic period house without living in it in an authentic way, and this is now very rarely acceptable. It is, however, honest and authentic enough to give a historic building modern amenities if it is part of a holistic attempt to keep an important building useful; and, without use, almost all of them will in time be lost. The true purpose of listing and the listed building consent procedure is to give the building itself a fair say in its own future. This surely is not too much to ask.

Endnotes

1 PPG15, p.1 (1.1).
2 PPG15, p.1 (1.3).
3 Loyd Grossman, 'Sharing the past with everyone: Engaging with England's heritage', *Conservation Bulletin* 50 (Autumn, 2005), 33–4.
4 Georgian Group Conference Notes, 1948.
5 English Heritage, *Heritage Counts: The state of England's historic environment* (English Heritage, London, 2005).
6 John Harvey, *Conservation of Buildings* (John Barker, London, 1972), p.180.
7 Terence Blacker, 'I live in a "horror conversion" – and here's why', *Independent*, 23 November 2005.
8 *Bath Life*, December 2005, 9–13.

4 Regeneration and the historic environment

Duncan McCallum

Introduction

The historic environment lies at the heart of so much physical regeneration. From one perspective, those conserving the historic environment have long been undertaking what we now call 'sustainable development' and 'regeneration', but these labels were not then in common use. The historic environment sector is now much more conscious of the wider implications of its work to protect our heritage and is keen to point out the social, economic and wider environmental benefits that often occur when the historic environment is conserved. While conservation for conservation's sake rightly remains a cornerstone of the work of the sector, the importance of regeneration work in bringing benefits to the historic environment is arguably of much greater significance; it touches on more people's lives, it affects the local economy and the places it creates become familiar, and hopefully enjoyed, local environments.

The government's view

Statements by government on the positive relationship between the historic environment and regeneration have been an encouragement to the historic environment sector. The Deputy Prime Minister John Prescott advised: 'Put people first, restore historic buildings, find space for the new and the bold . . . and people come back into declining city centres.'[1] Having experienced an economic and cultural revival, many of the major cities in England, particularly those in the north such as Liverpool, Manchester, Leeds and Newcastle, have been successfully pursuing that path for a number of years and are seeing exciting conversions of historic buildings alongside high-quality new buildings, which have made these areas attractive places in which to live, work and be entertained.

The government's *Response to ODPM Housing, Planning, Local Government and the Regions Committee Report on the Role of Historic Buildings in Urban Regeneration* (November 2004), for example, stated:

> . . . we agree wholeheartedly with the Committee that the historic environment has an important part to play in regeneration. The Government agrees with the Committee's findings that historic buildings have already provided a foundation

for the regeneration of many towns and cities and that heritage-led regeneration reinforces the sense of community pride, makes an important contribution to the local economy and acts as a catalyst for improvements to the wider area.'[2]

The Department for Culture, Media and Sport issued a consultation document in 2004, *Culture at the Heart of Regeneration*, which takes a wider look at the important part culture, which includes the historic environment, plays in regenerating our cities, towns and communities. Although the roots of this thinking go back a very long way, the government's statement *The Historic Environment: A force for our future*[3] formed the starting point for a new wave of interest in using heritage to make regeneration schemes more successful.

English Heritage has been working hard to understand, measure and promote the many positive benefits of regeneration that involves the historic environment. The organisation has clear evidence of many of those benefits and in 2005 published a short policy statement, *Regeneration and the Historic Environment: Heritage as a catalyst for better social and economic regeneration*. The government's push to try to make development of whatever kind more sustainable was set out clearly in Planning Policy Statement 1: *Delivering Sustainable Development*,[4] which seeks outcomes where social, economic and environmental objectives are jointly achieved over time – that is, trade-offs should be avoided. The document mainstreamed the historic environment rather than treating it as an add-on and it emphasised the importance of local distinctiveness and community involvement. The voluntary sector, too, demonstrated the important part it plays in delivering regeneration in Heritage Link's *The Heritage Dynamo: How the voluntary sector drives regeneration*,[5] by illustrating success stories from around the country.

As politicians and planners seek to create new 'sustainable communities' and try to work out how to avoid some of the mistakes of the past fifty years, they sometimes forget that there are many historic settlements or parts of settlements where communities have been living successfully for generations. These areas are mixed use, they are built to a high density, they are human-scale, they are built up over time and they have seen a continual, small-scale level of change rather than wholesale replacement. These are almost always successful areas, popular with residents and businesses alike as well as with visitors who enjoy the 'fine grain', the unexpected juxtapositions and the vitality that comes from diversity.

The English Heritage policy statement *Regeneration and the Historic Environment: Heritage as a catalyst for better social and economic regeneration* set out the holistic approach to regeneration now being taken by the organisation:

> Successful regeneration means bringing social, economic and environmental life back to an area. It transforms places, strengthens a community's self-image and re-creates viable, attractive places which encourage sustained inward investment. The historic environment is all around us, and includes landscapes, parks and other green spaces, historic streets, areas and buildings, and archaeological sites. Regeneration projects cannot ignore it.[6]

The document set out key reasons why the reuse of heritage assets is critical if development and redevelopment are to be sustainable. These are based on the experience gained by English Heritage's engagement in thousands of regeneration projects since its inception in 1984. The main argument is a very simple one: that it is better to use what is there than squander the effort and resource that went into the initial construction by replacing it with something else. This does not, of course, apply in all circumstances, but so many traditional building types are adaptable, and the sensitive adaptation, alteration and addition of existing buildings can offer the user the comforting reassurance of the familiar along with the excitement and the technological advantages of the new. The nine reasons were stated as follows:

1. **Reusing existing buildings is a simple way of achieving sustainability.** Recent research undertaken in the north-west of England by English Heritage found that, based on projections over thirty years, the cost of repairing a typical Victorian terraced house was between 40% and 60% cheaper (depending on the level of refurbishment) than replacing it with a new home.[7] Reusing buildings saves waste and reduces the need for new building materials. Demolition and construction account for 24% of the total annual waste produced in the UK.[8]
2. **Reusing buildings and adapting landscapes help reinforce a sense of place.** Investment in the historic buildings and streetscape of Brick Lane, East London, by English Heritage, the Heritage Lottery Fund and other partners has strengthened the area's distinctive identity. Its revitalisation has helped the growth of Brick Lane as a focus for Bengali festivals and cultural events.
3. **New large-scale developments risk losing the fine grain that characterises historic areas.** Great care is needed in undertaking new development in sensitive areas to avoid the wholesale amalgamation of plots, straightening of building lines, loss of incidental spaces, flattening of silhouettes, ironing-out of irregularities and reduction in the mix of uses which all help to integrate the new with the old. The £750 million Paradise Street development in Liverpool has been carefully designed to knit the new development into the townscape of the historic Ropewalks area, recreating some of the area's historic street pattern and reusing many of the vacant historic buildings.[9]
4. **Reused buildings can often be sold for a premium compared to a similar new-build property.** Many historic buildings are seen to be more desirable than their more recent equivalents. Historic residential properties, for example, often carry a premium. Research suggests pre-1919 houses are worth on average 20% more than equivalent more recent houses.[10] The Royal William Yard in Plymouth, a 7-hectare (17-acre) early nineteenth-century former victualling yard for the Royal

Navy, was taken over by the South West of England Regional Development Agency when it became redundant. The subsequent conversion of two of the buildings by developers Urban Splash was such a success that all the apartments were pre-sold in a single day.

5. **Restoring the historic environment creates jobs and helps underpin local economies.** Work by English Heritage in the *Heritage Dividend* demonstrates that initial heritage investment in heritage-led regeneration projects levers in significant amounts of other capital and helps to sustain and create jobs.[11] The heritage-led regeneration of the Jewellery Quarter in Birmingham ensured that this vibrant and historic quarter remains a thriving centre for the manufacture and retail of jewellery with 6000 people employed by 1500 separate businesses. Nationally, the shortage of workers in many craft skills demonstrates the potential for further growth in employment.

6. **An attractive environment can help to draw in external investment as well as sustaining existing businesses of all types, not just tourism-related.** The transformation of the redundant eighteenth-century Royal Dockyard in Chatham, as well as drawing in almost 2 million visitors, helped attract 100 businesses employing over 1000 people and had a positive impact on the local economy as a whole estimated at £20 million a year.[12] A study of the economic value of the heritage coastlines, National Parks and Areas of Outstanding Natural Beauty in the north-east demonstrated that through businesses and the effects of tourism these areas generated output of £700 million and support 14 000 jobs. The majority of businesses considered the quality of the landscape and the environment to be a factor in their performance.[13]

7. **The historic environment contributes to quality of life and enriches people's understanding of the diversity and changing nature of their community.** Regeneration has to have the support of local people, otherwise it is likely to fail. People are often immensely proud of their local heritage. A recent MORI poll in the north-east found that, after 'people and a sense of community', 'heritage and the built environment' was what gave the region its special character.[14] Many areas have a rich legacy which contributes to local identity and is an important local educational resource. Ironbridge Gorge, Coalbrookdale, Shropshire (Figure 4.1), for example, played a unique role in the development of the Industrial Revolution in the eighteenth century. Its decline in the twentieth century was reversed following its designation as a World Heritage Site in 1986. The area has a vibrant community and a wide range of businesses, shops and community services providing employment for 1500 people within the World Heritage Site. Recognition of Ironbridge's social, economic and environmental qualities has been at the heart of the area's continuing successful regeneration.

8. **Historic places are a powerful focus for community action.** The British Urban Regeneration Association (BURA), in an analysis of best practice in urban regeneration, concluded that 'historic buildings can act as focal points around which communities will rally and revive their

Figure 4.1 Ironbridge Gorge, Coalbrookdale, Shropshire: the World Heritage Site.

sense of civic pride' and that 'care should be taken not to destroy old buildings before their potential is realised'.[15] The transformation of the early nineteenth-century St John's Church in Hoxton, in the London Borough of Hackney, to include a nursery school, a community café, an employment project and a fitness centre as well as its continued use as a church has strengthened its role at the centre of its community without destroying its contribution as a high-quality architectural landmark.[16]

9. **The historic environment has an important place in local cultural activities.** Historic buildings, streets and parks are often key venues for local events. In Queen Square, Bristol, the removal of the inner ring road from the early eighteenth-century square and the redesign of the open space to reflect the original layout enabled a range of cultural events from outdoor cinema to concerts to take place. It also provides informal recreation space and a fitting setting for the surrounding historic buildings.

In all of this it is easy to forget that heritage is also highly valued for its own sake, just as we value a painting or a play, and English Heritage is currently working on research which seeks to measure that value in economic terms.

Complicating factors that make historic environment regeneration more difficult to achieve

Not all regeneration that involves the historic environment is necessarily good, nor does it always have wider spin-off benefits. There is a range of factors that can lead to poor-quality or short-lived regeneration. Examples include the following:

1. Schemes that tackle one-off problem buildings or areas in isolation will bring it or them into a better state of repair, but if the social and economic conditions are not right the improvements are unlikely to be sustained in the long term. Such buildings or areas risk falling back into disrepair a few years later, and then will need further investment to try to rescue them once again. This is not only economically wasteful, but is also very dispiriting for the local community who can come to associate regeneration with a quick fix that temporarily brightens up an area without tackling the underlying problems.

2. There can be problems with owners who are unrealistic about what can be achieved or who are unwilling to see change. Many historic buildings officers are of the view that problem buildings that become 'at risk' are almost always capable of an appropriate new use: that it is inaction on the part of the owner that causes the problem. Some of the reasons for inaction may be entirely understandable, such as lack of resources, but sadly there is sometimes a degree of irrationality that makes the problems so intractable.

3. Although a significant proportion of problems are caused by property owners, there are buildings where no use would be financially viable. Sometimes this is because the building itself is of such a specialist nature that when the original use ends there is no practical new use. This is particularly the case with structures – redundant military structures can be particularly hard to reuse – but if increasing numbers of below-ground public conveniences in our larger cities can be converted to viable new uses such as nightclubs or bars, then the cases where no new use is ever likely to be found are probably few and far between, at least in economically buoyant areas. However, where the sums simply do not add up the case for public subsidy can be made and English Heritage, the Heritage Lottery Fund and a range of other funding sources may be able to fill that gap.

4. A growing issue for the historic environment sector is that, as the spread of what is defined as 'heritage' becomes ever broader, the pressure on the scarce resources available to protect it becomes very much greater. The appreciation of the historic and cultural value of whole landscapes, with relatively few – perhaps just locally important – buildings, is welcome but adds to the challenge of appropriate

management. Regeneration necessarily involves change, and it is vital that the appreciation of the huge amorphous mass of what we regard as 'heritage' does not become a rigid dogma to protect every entity from change.

5. Another issue that makes decisions concerning historic environment and regeneration challenging is the fact that different parts of society and different communities will place different values on elements of the historic environment. We know that local people value heritage very highly; a public opinion poll[17] found that more than eight out of ten people agreed with the statement 'the heritage in my area is worth saving'. In Cornwall, an area associated with a very strong local identity, the figure was 91%, but even in Bradford and London the figures were 85% and 82% respectively. However, within a local community there will be widely differing views on what the local heritage actually is. Some may place more emphasis on a handful of landmark buildings, places or spaces, while others may value more modest elements such as areas of artisan housing, a canal system or modest industrial buildings that were critical to the economic development of the area. Another group may feel that their heritage is a less tangible thing and is tied up with the sights, smells and ambience of a place, and that the built fabric simply acts as a backdrop to a more dynamic view of heritage and culture. A successful regeneration scheme should seek to understand the variety of perceptions in a local area and reflect and perhaps seek to reinforce them in the final scheme.

The historic environment sector has come a long way in recent years in being able to counter the critics who accuse heritage of constraining development, of continuing to cause planning blight and decay, or causing uncertainty and delay, or being detached from reality and not understanding development economics. The push from mainstream planning for development to be more sustainable has met the historic environment sector halfway and there is much more common ground, but there are still many disagreements, particularly in areas where a strong economy is driving the pace of change very fast. Conversely, in those areas where the market is failing, there is conflict over the dramatic interventions that many feel are necessary to 'kick-start' a local economy. The most notable example of this is in relation to areas of low-demand housing, the nine Housing Pathfinders identified by the government for housing market restructuring. In these areas a significant amount of demolition of small terraced housing, along with refurbishment and new building, is proposed.

English Heritage set out its position in relation to the current round of proposals for the regeneration of low-demand housing;[18] it accepts that some demolition is necessary and the historic environment is not the only criterion for deciding which houses are to be demolished. It wishes to see an informed assessment of what is there, and its potential for adaptation, occurring before key decisions are taken. It believes that local communities as well as specialists should feed into this process. Complete clearance of an area is rarely justified; rather, change should be knitted into the existing environment to give real community continuity. In areas where the historic environment is distinctive, retains its coherence and is valued by the local community, English Heritage favours an approach which promotes repair and refurbishment as an alternative to outright replacement.

Understanding the role of heritage in regeneration

The phrase 'heritage-led regeneration' is often used by the historic environment sector to describe schemes where heritage has played a significant part in the regenerative process. However, the term is only really appropriate in situations where the scheme would not have happened but for a significant historic environment element. Such schemes are widely known, such as the highly successful regeneration of part of the centre of Newcastle upon Tyne, the mid-nineteenth century Grainger Town (Figure 4.2). Conceived and built as a planned city quarter between 1835 and 1842, Grainger Town was designated a conservation area containing 244 listed buildings, but in the late twentieth century it suffered major economic and social decline.

The Grainger Town Partnership was established to tackle this problem and at the end of March 2003 approximately £174 million had been attracted into the area, including £146 million from the private sector. The project is now widely recognised as an exemplary regeneration scheme involving private and public sector partnerships. Its main thoroughfare, Grey Street, was voted 'Best Loved Street in Britain' by CABE and BBC Radio 4 listeners. The large number of high-quality historic buildings within the regeneration area was pivotal in defining the end vision. Most

Figure 4.2 Grainger Town, Newcastle upon Tyne, a planned city quarter of 1835–42.

regeneration schemes that involve elements of historic environment do not start with such rich material.

Another example of heritage as the catalyst is the town of Frome, Somerset. Following economic decline over many years, efforts by the local community and the town, district and county councils with support from English Heritage, the Heritage Lottery Fund and others reversed the fortunes of this important medieval textile centre. The conversion of the derelict 'Feather Factory' in Willow Vale to residential units is just one of the many successful regeneration and enhancement projects which have given the town a new lease of life (Figure 4.3).

There is a spectrum of levels of interaction between the historic environment and regeneration projects. At one end there are schemes, such as the Nottingham Lace Market, where heritage is the catalyst for regeneration. In the middle is the situation where the historic environment is a quality component of a regeneration scheme, typically where some historic buildings are being retained but there is a significant element of new build to recreate a vibrant area. Liverpool Ropewalks exemplifies this level of integration. At the other end of the spectrum is the concept of heritage as a token. In many former mining areas in the Midlands and the north, there are few signs of the former industry, but a pit wheel may sometimes

Figure 4.3 The 'Feather Factory' in Willow Vale, Frome, Somerset.

be seen fixed in an ornamental brick support at the entrance to a new industrial estate. Pieces of industrial equipment, chimneys or other landmarks are sometimes retained but in these cases no real attempt is made to integrate the historic environment into the regeneration of the area in a way that keeps some of the historic context. Although the 'catalyst' schemes tend to be the ones that get the most publicity, the 'quality component' and, sadly, 'token' schemes make up the bulk of the regeneration schemes that happen in England. The goal for the sector should perhaps be to try to push all regeneration schemes further towards the 'catalyst' end of the spectrum.

A checklist for successful regeneration

English Heritage has had considerable experience of a wide range of regeneration schemes across the country. This experience suggests that there are some universal lessons for successful historic environment regeneration schemes:

1. **A strong vision for the future** – that inspires people and makes them want to get involved.
2. **A respect for local residents and businesses** – who have often fought hard to stop an area declining; ensuring they are included in a regeneration partnership means the project starts with community commitment.
3. **A tangible link to the past** – since places are not created in a vacuum, and people need familiar elements, visual reminders and a sense of continuity: landscapes, streets, spaces, buildings and archaeological sites play a part in defining a sense of place.
4. **An understanding of the area** – since knowing what exists and how it came to be as it is makes it easier to plan its future.
5. **A respect for what already exists** – making sure that places people will value are kept for the future.
6. **A record of the area before work starts** – so that future generations can understand how the site has evolved.
7. **An integrated, sustainable approach** – not concentrating on a particular social, economic or environmental consideration or a single use.
8. **Achieving the right pace** – regeneration that happens too quickly can harm the fabric and the community, while that which happens too slowly fails to create the momentum, commitment and enthusiasm needed to make a scheme a success.
9. **The highest quality design and materials** – to enhance local distinctiveness and sustain a sense of place that people can be proud of.
10. **Early discussions between the community, the local authority and other interested parties** – ensuring that options can be discussed and designs modified at an early stage, before too much has been committed.[19]

Conclusion

There are many lessons we can learn from the past about ways of achieving successful regeneration. There is now ample evidence that needlessly sweeping away buildings, places and spaces with real value is not good use of our scarce resources. Most places have the potential for imaginative renewal, and just as the historic environment we value today is made up of countless phases of development, so we should be confident in managing change today in ways that will combine the best of the past with the future.

Endnotes

1   *The Role of Historic Buildings in Urban Regeneration*, 11th Report of the ODPM Housing, Planning, Local Government and the Regions Committee Session 2003–4, 21 July 2004.

2   Office of the Deputy Prime Minister, *Government Response to ODPM Housing, Planning, Local Government and the Regions Committee Report on the Role of Historic Buildings in Urban Regeneration* (TSO, London, 2004).

3   Department for Culture, Media and Sport, *The Historic Environment: A force for our future* (DCMS, London, 2001).

4   Office of the Deputy Prime Minister, Planning Policy Statement 1: *Delivering Sustainable Development* (HMSO, London, 2005).

5   Heritage Link, *The Heritage Dynamo: How the voluntary sector drives regeneration* (Heritage Link, London, 2005).

6   www.helm.org.uk

7   English Heritage, *Heritage Counts* (English Heritage, 2003).

8   www.defra.gov.uk/environment/statistics/waste

9   Heritage Lottery Fund, *New Life: Heritage and regeneration* (Heritage Lottery Fund, London, 2004).

10 Nationwide Building Society, *What Adds Value* (2003).

11 English Heritage, *Heritage Dividend 2002* (English Heritage, London, 2002). In the conservation areas included in the study, £10 000 of heritage investment levered in £46 000 match-funding from private sector and public sources and delivered one new job, one safeguarded job, one improved home, 41 m^2 of improved commercial floor space and 103 m^2 of environmental improvements.

12 Heritage Lottery Fund, *New Life*.

13 ONE North East, *The Economic Value of the Protected Landscapes in the North East of England* (SQW Ltd, Cambridge, 2004).

14 English Heritage, *Heritage Counts 2004 (North East)* (English Heritage, 2004).

15 S. Burrows and P. Roberts, *Learning from Experience: The BURA guide to achieving effective and lasting regeneration* (British Urban Regeneration Association, London, 2002).

16 English Heritage, *Heritage Dividend*.

17 MORI poll for English Heritage (2003).

18 English Heritage, *Low Demand Housing and the Historic Environment* (English Heritage, London, 2005).

19 English Heritage, *Regeneration and the Historic Environment* (English Heritage, London, 2005).

5 Problems and opportunities in rural conservation

Jeremy Lake

The protection afforded to the built environment in rural areas has developed against the background of profound rural change. By the 1930s, when the environmental movement was blossoming into a diversity of lobbies – to promote access, stem ribbon development and restrict the march of the infrastructure for sustaining modern lifestyles across the landscape – the United Kingdom already had the lowest percentage of the working population in agriculture in Europe.[1] At that point in time, agriculture was in the grip of a long depression that had commenced with a run of poor harvests and cheap imports in the late 1870s and that ended with the boost to production forced on the state through war. New technologies and the restructuring of the agricultural industry have since seen the population of workers in agriculture dropping from 7% in the 1930s to 2% today. Attitudes towards this change have, however, long been riddled with challenges and paradoxes – early nineteenth-century romanticism, a visceral dislike of suburbanisation, indeed of the very infrastructure in the form of roads, electricity pylons and so forth that have sustained the diversification and viability of rural economies and communities, inward migration from urban areas, and the enjoyment of the countryside by the public at large. Perhaps the most prescient of the interwar commentators was Patrick Abercrombie, who in 1934 wrote that the countryside – a term increasingly coming into vogue as expressive of national patrimony, and an anchor of tranquillity and beauty in a rapidly changing world – is a complex blend of the 'natural' and the 'technological', not 'a Museum piece that can only be preserved and repaired as we treat a ruined abbey'.[2]

Seventy years on, and the profile of employment in rural areas is radically different, with broadband and other technological developments opening the positive potential for live–work, and the diversification of on-farm activities and non-agricultural employment; and on the negative side an increasingly elderly rural population, marooned as global developments exert an inexorable upward push on the costs of transport and energy supplies. Some 47% of all list entries lie in rural areas – in villages, hamlets and the wider countryside – and they display an enormous range in terms of their date, character, context and problems.[3] The vast majority are in private ownership. Our evaluation of them has transformed over the last 150 years, and countless numbers formerly condemned by housing commissions and rural authorities as unfit for habitation are highly sought after, command high values in the marketplace and of course are subject to the desires of an increasingly affluent and mobile middle class. For many other buildings in the wider countryside, however, the future is dependent on their finding a use other than that for which they were originally intended, and solutions lie not in consideration of their merits as historic buildings alone, but increasingly in their role in the wider landscape and the changing demography and structure of rural communities and economies.

Listing

The protection of buildings through listing was initiated in 1944 by the Town and Country Planning Act. Its earlier origins and development are discussed in Chapter 11,[4] but in summary the coverage of rural areas has been uneven; listing has progressed in stops and starts, the earliest surveys being conducted in the immediate post-war years of petrol rationing. Rural areas remained poorly covered – despite the initiation of the so-called National Resurvey in 1968 – until the Accelerated Resurvey commenced a parish-by-parish survey of rural areas in 1982. However, although this resulted in the addition of over 300 000 buildings to the statutory lists, challenging issues arose from the nature of the work, and indeed of the built environment itself:

- Inevitably, given variable resourcing and the rolling nature of the programme, the standards of the lists vary from the well-resourced programmes in Devon and Cornwall, which provide exhaustive inventories of external and internal features, to surveys with minimal descriptions focused simply on identifying structures that fulfil listing criteria.[5]
- No survey is a fait accompli; nor can it provide 'once-and-for-all' judgements about historic buildings. For example, the identification of early interiors, including those of a medieval date, can be hindered by roof heightening, rendering and external alterations. Some district councils, particularly in areas covered early on in the Accelerated Resurvey, have a hundred or more buildings added to the lists as a result of spot listing. The reactive nature of spot listing was focusing effort away from the need to inform and contribute to the wider understanding and management of historic buildings by owners, developers, local authorities, etc.
- Changing perceptions and understanding, moreover, result in shifting standards for determining the eligibility for listing, which heightens the need for much better links to be forged between academic understanding and the designation and daily management of historic buildings.[6]

The lack of guidance, and the need for frameworks for assessment attuned to different types of building caused a yawning gap to emerge between the largely ad hoc approach of the relatively young and developing historic building profession, and archaeological approaches to protection, which had been transformed since the publication of PPG16 in 1990.[7] This provided non-statutory advice in the form of planning guidance for all nationally important remains, whether scheduled as Ancient Monuments or not, and saw the consequent rapid growth of developer-funded archaeology. The clear message was that archaeology should be a material consideration in the planning process.

The thematic surveys

In the mid- to late 1990s, initial efforts were made to address these issues through thematic listing of particularly threatened building types, which emphasised the need for evaluation to broaden its scope beyond designation, and to inform the wider appreciation of our historic environment and its future sustainable management. The survey of barracks, for example, established a clear link between the clarity and transparency offered by the thematic approach, the options for reuse and the wider benefits thus brought to local economies and communities. It emphasised, at a time when many of these sites were being sold for development or reorganised for changing military requirements, that statutory protection can stimulate rather than constrain imaginative new development that responds to a sense of place, many of the sites affected by recommendations for protection being transformed from candidates for demolition into highly sought-after real estate. Sharing the results of the research that underpinned thematic surveys, through publication, was also a key output of the thematic listing process.[8]

Other studies, including those of farmsteads (see below) and chapels, evolved from research narrowly focused on designation to reveal a much broader complexity of issues. By the end of the 1980s, and with the Accelerated Resurvey largely completed, 150 Cornish chapels of all denominations (including 120 Methodist chapels) had been listed; the results had been strongly informed by the Royal Commission county inventories of chapels.[9] By 1993, the Methodist Church was expressing serious concerns to English Heritage about the selection process, as it appeared to be inconsistent and all the listings – at Grade II – excluded them from consideration for grant aid from English Heritage. It was also apparent that some of the finest examples had already been lost to conversion to domestic use, with the consequent total loss of internal fittings. The original intention to re-evaluate those chapels that had been listed broadened into a rapid survey of over 700 chapels and the compilation of data relating to the survival of chapel buildings and their fixtures and fittings, and then to consideration of their landscape and historic context. This provided a clear justification both for removal from the list, and for protection of key examples of the principal and best-preserved chapel types, the most significant being listed at Grade II*,[10] and also revealed the advantages of a fully contextual understanding.

What also emerged was a picture of the dynamism and flexibility of the Methodist movement and its remarkably varied architectural legacy, and the implications this has for the way in which we approach change driven by the needs of present and future chapel communities – without whom these buildings would generally be destined for far less sympathetic changes of use. Almost all chapels of the late eighteenth or the first half of the nineteenth century have been altered, have later fittings or have lost their fittings, thus exploding the hitherto common belief derived from earlier listing surveys and the Royal Commission county-based inventory that numerous early nineteenth-century chapels had retained their interiors. Very few retain virtually unaltered interiors or fittings; still fewer retain both – only 23 of over 700 chapels surveyed, for example, had retained all their box pews or benches. It was also clear that chapels exemplify through their enormous variation in size and architectural style the character and aspirations of their communities, and that they could not be evaluated through the lens of the Anglican Revival which even in the late twentieth century was exerting a grip on the way that architectural historians and others valued 'correct' Christian architecture – see Chapter 2.

The distribution of chapels also bore a strong relationship to patterns of settlement. The strongholds of Cornish Methodism, particularly in the rural industrial landscapes of the centre and west of the county, found no national parallels – with the notable exception of the mining valleys of south Wales – for the dominance that Methodism held, as a popular evangelical movement, over other forms of Christian worship. The message was clear – recognise rarity where it exists, but also ensure that this historical understanding of processes informs the work of present and future chapel communities and the distinctiveness of Cornish landscapes and culture as a whole (Figure 5.1).

The survey results were published, in partnership with the Methodist Church and the Cornwall Archaeological Unit, and launched a conference ('Chapels: A Bane or a Blessing?') in July 2001.[11] The conference concluded with an open discussion, from which various conclusions were drawn:

- Exteriors are of critical importance in retaining the historic character of all chapel buildings, and their contribution to local distinctiveness.
- The 'chapel heritage' embraces far more than the built environment – music, crafts and the concept of 'connexion'.
- Chapels have the potential to be central to the lives of rural communities, but in some areas there is inadequate community provision, and in others government funding has led to over-provision and 'by-passing' of chapels ideally suited for community use.
- Heritage professionals need to be mindful of the fundamental changes to organised Christian religion over the next generation, and the release of many more chapel buildings – especially those incapable of community use – onto the property market, and their distribution in relation to churches of other denominations (Anglican in particular).
- Preservation as found is only appropriate for a minute proportion of the total resource.[12]

Figure 5.1 Chapel of 1863 at Wheal Busy, Chacewater, one of only six wayside chapels that have retained all their original box pews (Eric Berry). The most complete examples of the small wayside chapels characteristic of small rural communities date not from the early nineteenth century, as was commonly supposed, but from the 1860s. Gems such as this have limited capacity for change, and it is vital that heritage professionals work with the communities and individuals that will be directly responsible for their conservation.

- Designation is widely regarded as burdensome if it is not supported by helpful guidance or grant aid for outstanding examples where compromises are forced on chapel communities.
- There is a widely held view, not shared by all members, that the demands of present-day mission and worship are incompatible with the layouts of earlier chapels.
- Chapel communities are vital to the long-term conservation of historic chapels, and there needs to be a better dialogue between planners and communities so that adaptation and extension can allow the main congregational space to be retained, respected and maintained.

This project – and work since conducted for the Anglican Church – highlighted the need for churches of all denominations to be evaluated in terms of factors such as community provision, and for approaches towards adaptation and reuse to be informed by an understanding of the pressures for change and their capacity to adapt and meet these needs.[13] The thematic surveys as a whole also heightened the need for academic information to be synthesised into more accessible language and integrated into the mainstream of popular and professional perceptions of the historic environment, as otherwise misconceptions would exist about what the built and wider historic environment can reveal about past economies and societies.

That the study of the built environment cannot, indeed, exist in a state of splendid isolation from other disciplines or the broader environment and communities was highlighted by other thematic projects, notably on military and industrial sites (see Chapter 6). These heightened awareness both that there was a need for a coordinated and multidisciplinary approach that removed the barriers that had hitherto existed between the recording of buildings and archaeology, and that their significance and character – as monuments, as buildings capable of adaptive reuse and as areas – conditioned the type of management that would best ensure a site's or structure's long-term future. The thematic study of airfields, originally focused on buildings, thus developed into a realisation that they not only had to be considered as functionally interdependent ensembles but that evaluation had to be linked to a wider understanding of their overall significance as places. As a consequence, key sites were identified for protection as conservation areas, which can be designated by local authorities, linked to character appraisals and in some instances management guidelines.[14]

This approach was taken forward into the study of Cold War sites,[15] although difficulties remained concerning the extent to which the tools at our disposal for the evaluation of buildings could be aligned with the much more strategic approaches towards archaeological resource management enabled by the publication of PPG16 and developed by the Monuments Protection Programme (MPP).[16] This work also raised challenging questions about how these sites could continue to change and be adapted, and indeed the fact that this does not simply represent the destruction of historic landscapes – 'Barbaric England of the scientists, the military men and the politicians' – but instead they must be understood and managed as an integral part of the process of change and creation in the landscape.[17] Over time their impact has softened; some have been reclaimed for agriculture, invaded by scrub or reworked into a variety of uses. Some are now important reservoirs of fauna and flora, such as the Royal Flying Corps' first squadron base at Netheravon in Wiltshire (founded 1912), and others are highly valued and interwoven with the lives of their surrounding communities.

Towards an integrated approach

Although thematic surveys helped to build better communications and confidence between all parties, there remained a feeling that approaches had to be more strategic and rounded in their scope. Designation – which at present covers 5% of the historic environment and 2% of the total built environment – had to be very tightly focused in order avoid weakening its effectiveness or its public support. Together with the work of MPP, they raised many of the questions that have since been tackled by the Heritage Protection Review, both in its proposals for a single unified register and the importance of developing Historic Environment Records on a statutory basis, and much broader considerations. Government, indeed, has stated that any reform of the system for heritage protection must engage with valuing the general character of the whole historic environment as well as designated highlights.[18] This is echoed by a wide variety of key players – business and community groups, developers, land managers and owners, as well as MORI polls – who assert that designation is not sufficiently rooted in its wider context, in terms of areas, communities and the forces for change, and that there needs to be a more integrated approach towards understanding the whole environment and engaging with local communities.[19] The issue is not simply one of speeding up planning applications or addressing the deficit of suitably qualified conservation officers, important as these are.[20] It is about getting the information that is needed into the working practices of those outside the historic environment sector, an issue that as we shall now see has been heightened by the challenges posed by the principles of sustainable development and the requirement for integrated delivery of environmental, social and economic objectives.[21]

This can be illustrated by looking at how attitudes towards the countryside as a whole have developed over the last sixty years. The Agriculture Act 1947 put in place the policy framework that anchored government subsidy and guidance to the restructuring and intensification of all aspects of the farming industry. The consequence for the farmed landscape – which still occupies 80% of the land area in the United Kingdom – has been profound, as has the impact on fauna and flora of hedgerow removal, chemicals and the loss of ecologically rich pasture and meadow. By the 1970s, awareness had developed concerning the trade distortions imposed by farm subsidies, along with growing realisation of the impact of production-based subsidies on wildlife, the environment and even the social and economic fabric of rural communities.[22] Environmentally Sensitive Areas were introduced in 1986, but despite their overall success concerns – expressed much earlier by thinkers such as Nan Fairbrother[23] – continued to mount regarding the long-term sustainability of designated sites and areas, and the pressures thereby exerted on their boundaries and on the 'white areas' in between. These informed the eventual adoption of more integrated approaches to land management, and the recommendations in the Curry Report of 2002 for the introduction of entry-level agri-environment schemes open to all farmers and of higher-level schemes focused on priority features and areas.[24] Although, however, the agri-environment schemes were adopted in this form in 2005 – and include a range of measures targeted at the historic as well as the natural environment, and fall within the bracket of 'green box' payments that are not considered to distort world trade – they are only one component of the so-called Pillar Two, which takes a meagre 3.5% of the budget of the EU's Common Agricultural Policy.[25]

New approaches towards the understanding and management of the environment, termed landscape character assessment, developed in response to these developments. These were initially focused on the identification of 'special landscapes', but over the 1990s developed into tools for mapping all landscapes and thus for informing change at a strategic and landscape scale above that of individual sites and designated areas. The best known of these tools in England is the array of 159 Joint Character Areas, each resulting from the mapping of a combination of factors such as land cover, geology, soils, topography and settlement, and field patterns, and extended by local authorities to a finer scale.[26] These Joint Character Areas are also being deployed as the framework for the delivery of advice and the targeting of resources through the agri-environment schemes, and for reporting change in the countryside in the Countryside Quality Counts (CQC) project.[27]

What was lacking from landscape character assessment, however, was an understanding of the 'historic' in the environment – rather than the 'historic environment', as if it is somehow separate from ecology and the physical landscape. There developed a growing awareness – revealed through a succession of publications – of the close interrelationship of historical, archaeological and ecological approaches towards past, present and future landscapes.[28] This awareness helped mainstream environmental and spatial planners, other disciplines and professions and local communities to understand what they have in the broader context. It provided them with the tools to understand the environment around them as a product of past change, and to understand what is important, and why.[29]

Two major projects, commissioned by English Heritage, thus expanded beyond their original intention simply to identify the most important landscapes and to evaluate types of settlement for designation purposes. The first of these projects, Roberts and Wrathmell's *Atlas of Rural Settlement in England* (2000), built on the work of previous generations of scholars and then took the patterning of settlement across the English landscape as the basis for analysis and the mapping of distinct areas that presented challenging questions about the historical development of settlement and land use.[30] The second project, and a complementary study to the first, is Historic Landscape Characterisation (HLC). HLC, as developed by English Heritage in partnership with county-based archaeological teams, is a rapid 'top-down' means of assessment that uses recent and historic mapping, and aerial photography, in order to convey how the landscape has evolved over time. It does this through identifying the present patterns of field boundaries, woodland and so forth in a given area, and the traces of earlier landscapes. Of critical importance in this respect is the use of geographical information systems (GIS) mapping technology, which enables the analysis of different sets of spatial data in order to record the key attributes that lead to definitions of various types of landscape, such as the 'reversed-S' curves indicative of medieval plough strips and the surveyor-drawn regular boundaries dating from the late seventeenth to the nineteenth century.

HLC and its urban cousin, the Extensive Urban Survey of small towns and the mapping of large conurbations,[31] work best at a high scale rather than at a detailed level, but this enables us to map the time-depth of large areas, and go beyond the points on the map that heritage data has traditionally comprised, and that have inhibited the historic environment sector from being more strategic in its scope. Work in rural and peri-urban areas aimed at informing the options for development in the Milton Keynes and Harlow–Stanstead areas, initially focused on identifying the spatial distribution of designated buildings and sites, have thus broadened into characterisation studies aimed at understanding the time-depth, character and sensitivity to change of the whole historic environment.[32] As a tool, available at the point of need, HLC is now informing a broad range of conservation and enhancement strategies, strategic land use planning, research agendas and similar initiatives, affirming the importance of human action, both past and present, to our perceptions of landscape and its physical and natural diversity of life and form.[33]

This is a question-based approach, broadly analogous to the way in which a building historian can appreciate a farmhouse or streetscape, for instance, and draw a distinction between the date and architectural style displayed by facades and the clues of earlier phases of historical development betrayed by plan form, the siting of stacks and other features. Field survey can also be personally and politically worrisome for owners and their representatives, and there will never be the considerable resources available to fund surveys at a comprehensive scale. Despite the work of many groups and individuals, very little information other than listed buildings records has found its way into county-based Historic Environment Records. There is little appreciation of the broader value of recording and archiving, and certainly little awareness of how it could respond to a high-level framework that can test and refine the conclusions reached, and thereby deliver better value to both the client and the community at large.[34] The consequences are ill-informed approaches to managing change and targeting resources, or to defining local distinctiveness.[35] Instead we must bring together what we do know, rather than be inhibited by awareness of what we do not, and present questions about the character, development and research potential of buildings at a landscape scale. GIS mapping indicates that such an approach can inform the way that we approach the study of the built environment (Figure 5.2).[36]

Although the concept of how to define and capture public values and feed these into the way that places and sites develop and change has also developed in recent years,[37] the historic environment sector has not been proactive in engaging with this issue. This is particularly true of buildings as opposed to landscapes, for – as Adrian Forty has pointed out in a masterly essay on post-war architecture – architectural historians have been more at home in describing 'the specifically architectonic themes of architecture' and the development of functional types than our 'physical perceptions of objects'.[38] Local communities attach huge importance to the built environment's role in providing them with a sense of place, but it has been the Countryside Agency rather than English Heritage that has administered the wide range of overlapping community-based initiatives – Vital Villages, Parish plans, village assessments and village design statements – whose key aim has been to relate existing features (settlement form and development, built environment) to the issue of acceptable and sustainable new development, and in the case of Parish plans to the identification of values, problems and opportunities.[39]

There is now, moreover, an increasing acceptance, embodied in the European Landscape Convention's definition of landscape as 'an area, as perceived by people, whose character is the result of the action and interaction of natural and/or human factors', which the UK government ratified

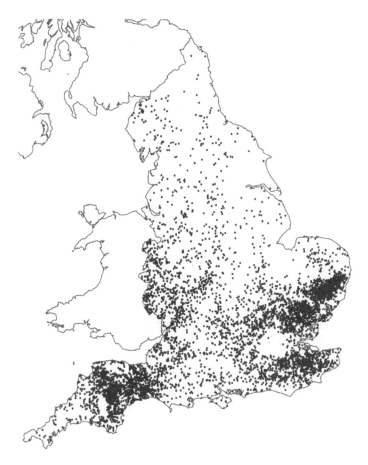

Figure 5.2 Map of pre-1550 farmhouses, indicating regional patterns of survival (© Crown copyright. All rights reserved. English Heritage 100019088. 2005).

in February 2006, of the critical role of public and professional perception as well as innate landscape, or architectural or historic character.[40] Indeed, once it is realised that value resides less in the thing itself than in the values that people attach to it – changing attitudes to Victorian architecture, to take one of many examples – the way is open to the exploration of new methodologies for linking the range of values applicable to a site or land-scape and to consideration of the options for change at the masterplanning stage. Public consultation, for example, informed the final drafting of a values paper and its incorporation into the masterplan and the conservation management plan for the famous code-breaking centre of Bletchley Park, in order to guide a flexible and area-based development framework avail-able at the point of need and thus provide owners, local authorities and potential investors with the confidence to plan for and invest in the site.[41] New characterisation tools being developed in urban contexts, such as in Lincoln, also have considerable potential to inform the future direction this work could take, and its capacity to influence planning, identify priority areas for research and reach a broader cross section of the community than is currently the case.[42]

It is significant, in this respect, that national planning policy – in parallel with its deployment of a less stringent approach towards rural development first observable from 1997 in PPG7 and then in the Rural White Paper – is placing more emphasis on both better-quality design and greater use of evidence-based and place-specific guidance and directions that are intended to complement the protection afforded by designation.[43] PPG 15, in 1994 (2.26), stated that local planning authorities 'should take account of the historic dimension of the landscape as a whole rather than concentrate on selected areas'. Thus Planning Policy Statement 1 (*Delivering Sustainable Development*) clearly states that 'Design which is inappropriate in its context, or which fails to take the opportunities available for improving the character and quality of an area and the way it functions, should not be accepted' (34) and PPS7 (*Sustainable Development in Rural Areas*) states that policies at a local level should include the 'need to preserve, or the desirability of preserving, buildings of historical or architectural importance or interest, *or which otherwise contribute to local character*' (PPS7, 19).

The need for high-quality design, informed by an understanding of local character and context, has been reinforced by the Department of Communities and Local Government's *Guidance on Changes to the Development Control System*, effective from August 2006, and related guidance by CABE. Applicants are now required to prepare design and access statements at the outset of a scheme, which are intended to demonstrate how the design process has been informed by a good understanding of local characteristics and circumstances. Complementary to these developments is the framework for the planning system introduced by the Planning and Compulsory Purchase Act 2004. Planning Policy Statement 11: *Regional Spatial Strategies* and Local Development Documents are now envisaged as the key means for maintaining and managing the environmental, economic and social value in rural areas. Local Development Documents will each contain a core strategy, supplementary planning documents, area-specific policies and action plans, and Statements of Community Involvement which are intended to enable local communities to understand and participate in the planning process. The latter contains recognition that regeneration will only be sustainable in the long term if it includes those local communities that play a critical role in the improvement of their own environments.

Farmsteads

We shall conclude by taking farmsteads as an example of how such approaches to rural conservation could develop. The wholesale redundancy of traditional farmstead buildings has been hastened by the post-1950 intensification and restructuring of the agricultural industry, the amalgamation of farm holdings and animal welfare standards. By the mid-1990s it was clear from buildings at risk surveys that historic farm buildings were the rural building type most at risk, from dereliction on the one hand

and from insensitive conversion on the other. Thematic listing surveys focused on the production of exemplar listings in consultation with owners and others were initially considered to be a means of addressing this issue, and Norfolk was selected as a pilot county in which to compare and contrast the statutory lists (compiled in the early 1980s) with the results of detailed survey work undertaken by the Centre for East Anglian Studies in 1986–7. Analysis of the lists highlighted the almost total dominance of barns to the detriment of other building types, and the inadequacy of guidance on building types and whole farmsteads representative of the development of regional farming traditions (Figure 5.3).[44]

It became increasingly apparent, however, that this was not a sustainable way forward. The process was too time-consuming for the resources available,[45] and was focused on the protection of a smaller and smaller number of buildings and farmstead groups, rather than engaging with the key issues of what listing was ultimately expected to achieve in the wider context of rural change, or informing how the whole resource of rural regeneration and the work of other agencies and organisations should be managed. The all-important context within which informed decisions concerning future designations and management should operate was lacking, as also was information on the drivers for change. Other concerns related to the perceived inadequacy of current policy. Designation excluded the vast majority of the total resource that can be defined as 'historic' or

Figure 5.3 Farmstead in the Forest of Arden, retaining two timber-framed barns and a stable; the farmhouse and other buildings were rebuilt from the late eighteenth century (Peter Gaskell). The combination of timber frame and eighteenth–nineteenth-century brick is a characteristic feature of the anciently enclosed landscapes of the West Midlands.

contributory to regional character and distinctiveness. Despite policies by English Heritage and local authorities designed to resist conversion to domestic use, it was increasingly realised that, within a planning system that since the Scott report of 1942[46] and the 1947 Town and Country Planning Act was designed to resist new development in the countryside, redundant listed buildings were of high financial value and actually attracted conversion, often of poor quality. Even the better-quality schemes could experience the 'death by a thousand cuts' of minor but cumulatively damaging works.[47] Concerns were also expressed by some local authority conservation officers over the practice of delisting converted farm buildings in cases where care had been taken to negotiate a design solution which they felt preserved the fabric and character of significant buildings. It was realised that a stress on aesthetic rather than 'archaeological' and landscape considerations in the delisting process was in danger of undermining the resolve of local authorities to expend resources in seeking acceptable outcomes.

Such concerns led to the Historic Farmsteads: Audit and Evaluation project, commissioned from the University of Gloucestershire by English Heritage and the Countryside Agency. This has provided a valuable insight into the character of the listed resource, the pressures driving upon it and the effectiveness of current policy.[48] Not only are dramatic declines in the numbers of farm buildings expected over the next ten to twenty years, but one-third of farmers plan to convert their buildings and two-thirds plan to sell off farm assets. Some 57% of farmstead sites have been subject to planning applications since 1980, and nine out of ten applications for works to domestic listed buildings (farmhouses and farm cottages) were in fact for works to working farm buildings within their curtilages. Global pressures on farming – which now contributes less than 1% to gross national product – will only increase in the next few years, particularly in upland areas. It was also clear that, despite policies that encouraged economic use, the overwhelming majority of consents granted for working buildings were for conversion into permanent residential buildings (71%) and only 15% for business use. Photographic evidence has also suggested that by 2004 over 30% of listed buildings had been converted, almost all to residential use, and that while the rate of conversion was reduced in National Parks it was little different from the national average in Areas of Outstanding Natural Beauty and was especially strong in accessible urban areas.[49] This suggested that while a rigorous policy framework was vital, it needed to place far greater stress on quality rather than simply considering the type of conversion, and to be informed by an understanding of local and regional variety – of buildings, landscapes and the drivers for change. A number of case studies, which measured the attitudes of key players in the conversion process, found that, despite a general appreciation of the landscape and historic value of farmstead buildings, limited knowledge and guidance was leading to confused and often conflicting approaches to the evaluation of character, significance and capacity for change.[50]

A revised policy on traditional farm buildings published in 2006 stressed the likelihood of large-scale departures from the farming industry and consequent high levels of farm building redundancy, and that successful strategies for reuse must align an understanding of the regional and local character – in patterns of settlement and landscape, and in farmstead and building types – with their sensitivity to change.[51] Detailed guidance on the adaptive reuse of farm buildings has also emphasised the quality of design, both traditional and contemporary, including appropriate detailing, materials, craftsmanship and the setting of buildings.[52]

The audit project, however, also found that the majority of regional and local planning guidance, while addressing the issue of reuse, quotes from national guidance but reflects limited knowledge of the nature and character of historic farmsteads, whether at a local scale or in their broader context – partly because local authorities have felt vulnerable to challenge at appeal if they depart from national planning guidance. Limited knowledge of historic farmsteads in their broader context, and the lack of a consistent framework for understanding and valuing farmsteads and their buildings, was identified as the greatest obstacle to informing consideration of issues and potential difficulties at the outset of a project and ideally at the pre-application stage; to the targeting of priority features and areas for grant aid; and to the development of local plan policies that draw upon an understanding of historic farmsteads in their local and broader context.[53] Various stakeholders,[54] furthermore, have expressed the need for a product that

- is easy to use and update, and that engages with the value of the unlisted as well as the listed buildings stock
- identifies key farmstead and building types in their landscape, regional and national context
- builds on an understanding of character and context in order to guide the identification of priorities and the targeting and monitoring of grant aid
- highlights priority areas for research and monitoring, conservation, restoration or enhancement
- enables users, on the point of need, to make informed decisions about the priorities for grant aid and the options for the sustainable reuse of rural buildings.

It is clear, therefore, that we need to design and demonstrate new consistent and transparent methods that can inform the development of local plan policies and guidance for rural buildings that are based on a clear understanding of character and context and work from broad principles to detail. In order to do this, we must paint a picture based on what we know (rather than being, as experts, inhibited by awareness of the gaps in our knowledge), through producing broad, general statements about the character of the historic built environment at a landscape scale. An initial step comprised eight preliminary regional character statements which supported the revised policy, targeted at a broad diversity of users with an interest in researching, understanding and managing historic farmsteads. This represented an initial attempt to understand the farmsteads of each region in their national and landscape context.[55]

Pilot work in Hampshire and elsewhere has also explored methods for mapping the patterning of time-depth and farmstead character across the landscape. As a first step, descriptions relating to each of the Joint Character Areas and Hampshire's own landscape character areas and types were compiled. These outlined the character and landscape context of historic farmstead types and buildings, identified those features or elements that contribute to local distinctiveness and countryside character, and produced guidance and positive recommendations for enhancement based on this understanding.[56] These statements were further developed through consultation, by reference to the county Historic Environment Record and by rapid field survey. Information about farmsteads was captured by plotting all farmstead sites – not just those with listed or recorded buildings – as a separate map layer in GIS so that they could be analysed in relation to landscape character and historic landscape character (HLC) areas: the farmsteads entries on Hampshire's Archaeology and Historic Buildings Record were more than doubled to over 5300. This has demonstrated that the dating and distribution of farmsteads in the landscape, and the rates of survival of different types of steading and building, are closely related to patterns of landscape character and type.[57] The results of this work are being used to develop character-based supplementary planning documents and guidance for a range of applications including the targeting of priority features and areas, whole farm planning, integrated land management and research frameworks (Figures. 5.4 and 5.5).[58]

Figure 5.4 Saddlescombe in the Sussex Downs (Bob Edwards). Evaluation of this whole farmstead in its local and regional context, using the results of mapping over 15 000 farmsteads in an area of the south-east, has demonstrated that it is a uniquely well-preserved example of a farmstead of this type in the southern English chalklands.

Figure 5.5 An earth-walled stable/cow house typical of a New Forest commoner (Bob Edwards). The farm buildings of commoners in heathland areas offer little opportunity for reuse and so are highly vulnerable, but there is little information available. Most of these buildings are of relatively poor quality and are rarely listed. The mapping of farmsteads has highlighted the need to research this building type.

This work will help understanding of which features and areas have the greatest sensitivity to change, and thus should be targeted for public funding. In the Yorkshire Dales, for example, the decision was taken during the resurvey that, while the larger eighteenth-century and earlier field barns of the Craven Dales fulfilled listing criteria, the abundant, smaller and predominantly nineteenth-century field barns of the northern gritstone dales did not. The decision was taken by English Heritage and the Yorkshire Dales National Park to designate selected dales landscapes as conservation areas, with grants contributing up to 80% of the costs of repairs to walls and barns. This was in so many respects an admirable and brave decision, but subsequent auditing of the effectiveness of the scheme highlighted the fact that without a use they would either deteriorate or – with funding from agri-environment schemes (specifically the Environmentally Sensitive Areas scheme) – simply remain as iconic structures within highly valued landscapes whose amenity value contributed to the broader rural economy (Figure 5.6).[59] In recent years, as bale silage replaces hay as the fodder crop and the requirement to loose-house stock close to the main steading increases, it has become increasingly difficult to find a use for these buildings. Historic buildings are clearly expensive to maintain, and in the period from 2000 to 2004 they absorbed 40% of the £90 million

Figure 5.6 Small field barns that housed cattle over the long northern winters at Gunnerside Bottoms in Upper Swaledale, Yorkshire Dales (Jeremy Lake). Dairying had emerged as a major industry by the sixteenth century and the need to manage cattle and the valuable hay crop was a major factor in the enclosure of this landscape, where farmhouses were concentrated within settlements and were located in the fields. In the northern dales of Swaledale and Wensleydale, these barns make a major contribution to the character of the landscape. They were rebuilt as storeyed structures from the later eighteenth century, although traces of smaller and earlier heather-thatched buildings remain. Adaptive reuse, as here in Upper Swaledale, would have a major and undesirable impact on the landscape. Such landscapes, however, attract tourism and generate wealth for these areas.

spent on the historic environment under agri-environment schemes. Recent research has, however, established the benefits delivered to local economies and communities through these schemes in terms of both employment and the development of skills, and the importance of farmsteads within the context of highly valued landscapes. Of the 655 buildings in the Lake District ESA grant-aided between 1998 and 2004, 92% are now in productive use, contributing to improved efficiency in farm businesses and the generation of between £8.5 million and £13.1 million in the local economy.[60] Only thirty-five of these buildings are listed, a fact that underlines the importance of using landscape as a key aspect of any framework for assessment.

We must also understand the capacity of distinct farmstead and building types and their landscapes to absorb change, as recent work has shown that the adaptation of the existing building stock is accounting for as much housing growth in rural areas as in urban areas.[61] Some of the highest densities of historic farmsteads and pre-seventeenth-century buildings are concentrated within landscapes defined by dispersed farmsteads and hamlets and ancient patterns of fields and boundaries, such as in the High Weald of Sussex and Kent.[62] These areas have historically absorbed a great deal of change and are rich in biodiversity, but our understanding of such landscapes is in its infancy. The consequences, however, of planning policies that have forced development into settlement cores is both the erosion of historic villages and their boundaries and a lack of recognition of the long-term sustainability of communities living in landscapes characterised by isolated farmsteads and hamlets.[63] It follows that understanding the proportion of residences that are also used for business purposes, and thereby contribute to local economies, should be a research objective; this formed one of the major conclusions of a study conducted in Friesland in the mid-1990s, which found that in over 90% of cases the reuse of farmsteads for non-agricultural purposes was combined with supplementary income that contributed to local employment and the broader rural economy, but that this was only rarely in the form of explicit or primary sources of income that would be identified through postal address or land use data.[64]

This understanding of the historical patterning of the building stock, settlement and landscapes, combined with evolving patterns of live–work, will challenge some existing attitudes and policies but must inform an open debate about the future shape of our rural landscapes and communities. An assessment framework is being developed in consultation with land managers, planners and other key partners, aimed at *applying* the understanding of character and sensitivity to change to identify the options for change most appropriate to individual steadings or buildings. It outlines how the options for change can be informed by an understanding of character and context, and from this the site's sensitivity to change, working from the buildings' landscape setting towards the farmstead as a whole, and finally to individual buildings and their component parts (Figure 5.7 and the appendix to this chapter).

Conclusion

All this in essence boils down to two very simple messages. The first is the importance of working from a landscape scale as a framework for understanding buildings and how they developed, and for understanding their present and likely future state and how to manage change. The second is that the work of the historic environment sector must underpin the strategies and daily work of other disciplines and organisations that have a massive impact on the direction that future change will take.

Seeing buildings as part of landscape, and within their wider geographical and thematic contexts, can thus allow us to inform the options for change to our rural landscapes at a strategic level. This represents a shift in focus away from individual buildings to a more question-based and

Figure 5.7 The hamlet of Drebley in Upper Wharfedale, on the Bolton Abbey Estate, developed on the site of a medieval hunting lodge (Jeremy Lake). The boundary walls around the hamlet relate to the enclosure of the landscape from the medieval period, including the former communal arable in the foreground. It retains two cruck-framed and formerly heather-thatched barns, the gable end of one being visible to the left, which are very rare survivals of buildings once found in great numbers across the northern uplands until the nineteenth century. The three large combination barns, with areas for threshing and storing the harvested corn crop above accommodation for livestock, are typical of the larger-capacity structures which had become the dominant type by the mid-nineteenth century. Upland landscapes are again poised for considerable change in the future, and all these buildings are now redundant for the purposes for which they were intended. They also have markedly different capacities for change, determined by factors such as their physical form, their siting in the landscape and their rarity. Estates need to determine what buildings should be retained, adapted or salvaged for their materials.

holistic approach, and one that uses landscape to both reflect and inform the patterning of the built environment. Our aspirations for the historic environment also need to be set against a clear and wide-ranging understanding of the drivers for change. This must inform an open debate about the types of landscapes that we can envisage in the century to come. We need to look at the whole, to recognise that the *cultural landscape* is all around us, part of our daily experiences, subject to our perceptions of its value and the demands we place upon it through the lifestyles that we all crave. This is leading to a shift in the way that we use and manage buildings, not as objects in themselves but as an integral part of the living landscape, changing and adapting into the future.

Endnotes

1   G. Fairclough, 'Europe's landscape: archaeology, sustainability and agriculture', in G. Fairclough and S. Rippon (eds), *Europe's Cultural Landscape: Archaeologists and the management of change* (Europae Archaeologiae Consilium, Brussels, 2002), pp. 8–9.

2   Patrick Abercrombie, *Country Planning and Landscape Design: The Stevenson Lecture for 1933.* (University Press of Liverpool and Hodder & Stoughton, London and Liverpool, 1933), pp. 5–6, 27. For a useful summary of policies and development, see J. Sheail, *An Environmental History of Twentieth-Century Britain* (Palgrave Macmillan, Basingstoke and New York, 2002).

3   *Heritage Counts: The state of England's historic environment* (English Heritage, Swindon, 2005), p. 9.

4   There are also useful essays in *Transactions of the Ancient Monuments Society* 37 (1993), including essays by J.H. Harvey, 'The origin of listed buildings', pp. 1–20, and M. Robertson, 'Listed buildings: the National Resurvey of England: background', pp. 22–38.

5   Some parts of the country were not even covered during the Accelerated Resurvey. In Suffolk, for example, the lists were drawn up between 1983 and 1987 although the lists for the former Rural District Councils of Clare, Melford and Cosford in the south of the county date from the 1970s. Large swathes of the south-east have similarly not been surveyed since the early 1970s.

6   As outlined in PPG15: *Planning and the Historic Environment* (DoE/DNH 1994), 6.10–16.

7   PPG16: *Archaeology and Planning* (DoE, 1990).

8   See James Douet, *British Barracks: Their architecture and role in society* (English Heritage, London, 1998), pp. xvii, 198. Other thematic projects were undertaken in collaboration with the Royal Commission on the Historical Monuments of England (RCHME) – for example, J. Cattell and B. Hawkins, *The Birmingham Jewellery Quarter: An introduction and guide* (English Heritage, London, 2000).

9   C. Stell, *An Inventory of Nonconformist Chapels and Meeting-Houses in South-West England* (HMSO, London, 1991). This was one of a series of county-based inventories produced by the RCHME.

10 The survey resulted in the removal of twenty-eight chapels from the statutory list, the listing of thirteen at Grade II and three at Grade II*, and the upgrading of twenty chapels to Grade II*.

11 J. Lake, J. Cox and E. Berry, *Diversity and Vitality: The Methodist and Nonconformist chapels of Cornwall* (English Heritage/Methodist Church, Truro, 2001).

12 Such as Penrose in St Ervan, which was built in 1863 and survives as one of the most complete wayside chapels in Cornwall. This small gem is now in the care of the Historic Chapels Trust, who took a keen interest in the survey.

13 For work on Anglican churches – which constitute nearly half of all Grade I buildings in England – see T. Cooper, *How do We Keep our Parish Churches?*, Report for the Ecclesiological Society (Ecclesiological Society, London, 2006); also *Building Faith in our Future*, a policy document published in 2004 by the Church Heritage Forum on behalf of the Church of England, and a report by the University of Gloucestershire's Countryside and Community Research Group for the Gloucester Diocesan Rural Group – C. Short and R. Stickland, *A Vibrant Church: A report of the Church of England in rural Gloucestershire* (CCRU, Cheltenham, 2003).

14 The approach to twentieth-century military sites is summarised in C. Dobinson, J. Lake and J. Schofield, 'Monuments of war: defining England's twentieth-century defence heritage', *Antiquity* 71 (1997), 288–99; also English Heritage, *Monuments of War: The evaluation, recording and management of twentieth-century military sites* (English Heritage, London, 1998; 2nd edn 2003); for airfields see English

Heritage, *Historic Military Aviation Sites: Conservation management guidance* (English Heritage, London, 2003) and the essays in B. Hawkins, G. Lechner and P. Smith, *Historic Airports,* Proceedings of the International 'L'Europe de l'Air' Conferences on Aviation Architecture (London, 2005) including J. Lake, C. Dobinson and P. Francis, 'The evaluation of military aviation sites and structures in England', pp. 23–34. Work on extensive sites such as airfields, and on post-war buildings, informed the drafting of another key document commissioned by English Heritage and the Office for the Deputy Prime Minister in the period before the Heritage Protection Review – *Streamlining Listed Building Consent: Lessons from the use of management agreements* (English Heritage, London, 2003).

15 W.D. Cocroft and R.J.C. Thomas, *Cold War: Building for nuclear confrontation, 1946–89* (English Heritage, London, 2003).

16 For MPP, see D. Stocker, 'Industrial archaeology and the Monuments Protection Programme in England', in M. Palmer and P. Neaverson (eds), *Managing the Industrial Heritage: Its identification, recording and management*, Leicester Archaeology Monographs No. 2 (Leicester, 1995).

17 W.G. Hoskins, *The Making of the English Landscape* (Hodder & Stoughton, London, 1955), p. 299; A. Bradley, V. Buchli, G. Fairclough, D. Hicks, J. Miller and J. Schofield, *Change and Creation: Historic landscape character, 1950–2000* (London, 2004).

18 Department for Culture, Media and Sport, *Review of Heritage Protection: The way forward* (DCMS, London, 2004), pp. 4–5. This is true of other forms of designation as well. In the Yorkshire Dales, for example, the dominance of pastoral farming economies since the fourteenth century has conserved a diversity of evidence for past farming systems and settlement dating from the Bronze Age that is still visible. They are nationally important but far too extensive to be sustainably managed through scheduling, and they are on farmland that is exempted from control through PPG16. Existing designations, moreover, are overlapping and disparate, and can be hard for landowners to comprehend – remains of the lead mines on the moorland tops protected through scheduling, extensive areas of moorland protected as Sites of Special Scientific Interest, and conservation area designation relating to discrete areas of barns and walls but seldom to the settlements that contain within them many of the former farmsteads that managed all of this land.

19 These issues were addressed by English Heritage in *Power of Place* (English Heritage, London, 2000), by government in its response, *A force for our future* (DCMS/DTLR, 2001), and in spring 2005 to the DCMS Parliamentary Select Committee on the Heritage White Paper by a range of bodies including the National Trust, the Country Land and Business Association and others.

20 As examined by Oxford Brookes University in its reports on local authority practice and PPG15*: Information and effectiveness* (Oxford Brookes University, Oxford, 2000) and *Conservation Provision in England* (Oxford Brookes University, Oxford, 2004).

21 For more on the principles of an integrated approach towards sustainable development, see Countryside Commission, English Heritage and English Nature, *What Matters and Why: Environmental capital – a new approach* (Cheltenham, 1997); Countryside Agency, English Heritage, English Nature and the Environment Agency, *Quality of Life Capital: Managing environmental, social and economic benefit* (Cheltenham, 2001). The UK government definition of this can be found at www.auditcommission.gov.uk/qualityoflife/index.asp?page=index.asp&area=hplink (accessed 07/01/2007). For other work by the New Economics Foundation, the Sustainable Development Commission and others, see www.sustainablepss.org/appendix/appendixD.htm (accessed 07/01/2007) and the Strategy Unit report *Life Satisfaction: The state of knowledge and implications for government* (Cabinet Office, London, 2002).

22 See, for example, Countryside Agency, *Agricultural Landscapes: 33 years of change* (Countryside Agency, Cheltenham, 2006). Obtainable at www.countryside.gov.uk/ Publications/articles/Publication_tcm2-29715.asp (accessed 07/01/2007).

23 N. Fairbrother, *New Lives, New Landscapes* (Architectural Press, London, 1970).

24 Department for Environment, Food and Rural Affairs, *Farming and Food: A sustainable future*, Report of the Policy Commission on the Future of Farming and Food (Defra, London, 2002).

25 Entry Level Stewardship is a five-year agreement which covers the maintenance of traditional buildings. Higher Level Stewardship is targeted at high-priority situations and areas, and farmers applying for HLS can qualify for at least 80% capital grants. Despite the introduction of the Single Farm Payment in 2005, which decoupled subsidy from production, the latest round of the World Trade Organization talks have indeed failed to equate the issue of farm support with world trade. It is probable, therefore, that the future development of rural landscapes will reflect both the move to align UK agriculture with world markets, and the desire to conserve and enhance those landscapes that are more suited to smaller-scale, diverse and high-value production. For details of the scheme see *Entry Level Stewardship Handbook*, available at www.defra.gov.uk

26 See www.countryside.gov.uk/lar/landscape (accessed 07/01/2007).

27 The CQC analysis process is based on an assessment of change in the countryside relative to the *descriptions* written for the 159 Joint Character Areas (JCAs) in the mid- to late 1990s. A pilot project carried out across Hampshire in January 2004 demonstrated that it was possible to go beyond a single 'historical features' topic heading, as used in the Character Area Profiles for the 1990–98 CQC analysis, and produce draft historic profiles for each of the JCAs. See www.cqc.org.uk.

28 See Oliver Rackham, *The History of the English Countryside* (Dent, London, 1986); also, for perspectives from ALGAO (Association of Local Goverment Archaeological Officers) on archaeological resource management, see A.R. Berry and I.W. Brown (eds), *Managing Ancient Monuments: An integrated approach* (Clwyd Archaeological Service, Mold, 1995). For further reading, albeit on a subject that is changing rapidly, see J. Grenville (ed), *Managing the Rural Historic Environment* (Routledge, London, 1999).

29 For example, *Conservation Issues in Local Plans* (English Heritage, Countryside Commission and English Nature, London, 1996); *Sustaining the Historic Environment: New perspectives on the future* (English Heritage, London, 1997).

30 B.K. Roberts and S. Wrathmell, *An Atlas of Rural Settlement in England* (London, 2000); and *Region and Place: A study of English rural settlement* (English Heritage, London, 2002).

31 R.M. Thomas, 'Mapping the towns: English Heritage's Urban Survey and Characterisation Programme', *Landscapes* **7**,1 (2005), 68–92.

32 Went *et al.*, 'Strategic development, sustainable communities', *Conservation Bulletin*, Issue 47 (2005), 4–10.

33 For more information on characterisation as developed in English Heritage, see *Conservation Bulletin*, Issue 47 (2005), and J. Clark, J. Darlington and G. Fairclough, *Using Historic Landscape Characterisation* (English Heritage and Lancashire County Council, Swindon, 2004); S. Rippon, with J. Clark, *Historic Landscape Analysis: Deciphering the countryside* (Council for British Archaeology, York, 2004), pp. 100–142. Also see http://www.englishheritage.org.uk/characterisation (accessed 07/01/2007).

34 B. Edwards, 'Buildings archaeology in England: are the foundations in place?', in G. Malm (ed.), *Archaeology and Buildings* (BAR International Series 930, 2001), pp. 19–24; S. Orr, 'West Berkshire historic farm buildings: an assessment of the resource and guidelines for management', *Historic Farm Buildings Group Review* **4** (Spring 2006), 4–8; S. Gould, 'Analysis and recording of historic buildings within the

English planning framework: an assessment of current practice', *Archaeologist* **55** (2005), 12–13.

35 Staddle barns, for example, are targeted in the New Forest, the only problem being that there are none in this area. They begin to occur just to the north, in Cranborne Chase.

36 The GIS mapping of listed buildings poses important questions for future research. The concentrations on this map in mid-Devon, the southern half of the West Midlands and in particular the south-east and southern East Anglia do more than simply reflect our understanding of the distribution of taxable population and wealth in England in the period up to the early sixteenth century, when (with the notable exception of Cornwall) the most prosperous areas were concentrated south of a line from the Bristol Channel to the Wash. They are all areas marked by high to extremely high rates of dispersed settlement, where farmsteads were either isolated or grouped in hamlets and surrounded by landscapes of early enclosure that were established by the fourteenth century and subsequently adapted to varying degrees. The principal concentration of surviving early buildings – in a well-preserved and anciently enclosed landscape – north of this line runs from the Shropshire Hills to Herefordshire. Pre-1550 buildings are generally much sparser in those areas of central southern England where settlement in the medieval period was dominated by nucleated villages and extensive communally farmed fields, and where patterns of wealth were less evenly spread and more hierarchical in structure.

37 See Marta de la Torre (ed.), *Assessing the Values of Cultural Heritage: Research report* (Getty Conservation Institute, Los Angeles, 2002) and K. Clark (ed.), *Capturing the Public Value of Heritage*, Proceedings of the London Conference (English Heritage, Swindon, 2006).

38 A. Forty, 'Being or nothingness: private experience and public architecture in postwar Britain', *Architectural History* **38** (1995), 33–5. For more than a decade, in contrast, archaeologists have recognised the role that human experience and cultural perception have in our understanding of the physical remains of the past – see C. Tilley, *A Phenomenology of Landscape* (Berg, Oxford, 1994).

39 An example would be the Kent Downs AONB, where Village Design Statements have been used to communicate the views of local communities to planners. Information on Village Design Statements and other community planning tools is obtainable at www.countryside.gov.uk/LAR/Landscape/PP/planning/communities. asp (accessed 07/01/2007). Another example would be the hugely successful Local Heritage Initiatives, funded through the National Lottery. The Council for the Protection of Rural England's report (*Unlocking the Landscape*) has produced useful guidance on how local communities can engage in landscape character assessment.

40 G. Fairclough, 'Archaeologists and the European landscape convention', in Fairclough and Rippon, *Europe's Cultural Landscape*, pp. 25–37.

41 J. Lake, L. Monckton and K. Morrison, 'Interpreting Bletchley Park' in J. Schofield, A. Klausmeier, and L. Purbrick (eds), *Re-mapping the Field: New approaches in conflict archaeology* (Westkreuz-Verlag, Bonn, 2006); for the values paper, see www/english-heritage.org.uk/characterisation and for the masterplan, see www. mkweb.co.uk/urban-design/documents (accessed 07/01/2007).

42 David Stocker (ed), *The City by the Pool: Assessing the archaeology of the City of Lincoln* (City of Lincoln Council and English Heritage, Oxford, 2003). Rather than produce a research agenda for the whole of Lincoln, based on its Urban Archaeological database (the equivalent of the Historic Environment Record), the Lincoln Archaeological Research Assessment divided the city into individual Research Agenda Zones, linked to distinct eras in its development, which present questions to inform investigation of the city's past and issues to inform the future strategic and practical direction of work undertaken by the City Planning Service and the wider community.

43 Department of the Environment, PPG7: *The Countryside: Environmental quality and economic and social development* (DoE, London, 1997); Department for the Environment, Transport and the Regions, Rural White Paper, *Our Countryside: The future – a fair deal for rural England* (London, 2000).

44 Some 540 out of 590 listed farm buildings in 1997 being barns. The thematic survey of Norfolk farmsteads and the more general leaflet *Understanding Listing: The East Anglian farm* were both produced by English Heritage's Listing Team in 1997. See J. Lake and S. Wade Martins, 'Thematic listing surveys: the Norfolk pilot project', *Journal of the Historic Farm Buildings Group* **11** (1997), 1–16. Historic Scotland produced guidance for the listing of farmstead buildings, in order to inform its resurvey, in the same year – *Farm Architecture: The listing of farm buildings* (Historic Scotland, Edinburgh, 1997).

45 It should perhaps be noted that at the time when thematic listing was developing and dealing with a vast range of types from post-war buildings to textile mills there were only five members on the Listing Team, including the present author, who handled over 3000 spot-listing cases per annum and supervised a mostly urban rump of over 150 listing surveys.

46 Scott Report, *Report of the Committee on Land Utilisation in Rural Areas* (Ministry of Works and Planning, London, 1942).

47 As PPG15, para. 3.13, stresses, 'minor works of indifferent quality, which may seem individually of little importance, can cumulatively be very destructive of a building's special interest'.

48 P. Gaskell and S. Owen, *Historic Farm Buildings: Constructing the evidence base* (English Heritage and the Countryside Agency, London, 2005).

49 P. Gaskell and M. Clark, *Historic Farm Buildings Photo System: Data analysis,* research report for English Heritage (University of Gloucestershire, Cheltenham, 2005).

50 Basic definitions, and tensions between defining character and special interest, are not helped by the drafting of the 1990 Planning (Listed Buildings and Conservation Areas) Act. This states that listed building consent is required for 'any works for the demolition of a listed building or for its alteration or extension in any manner which would affect its character as a building of special architectural or historic interest' (Planning Act 1990, s. 7). However, section 16 (2) omits character and states: 'In considering whether to grant listed building consent for any works, the local planning authority or the Secretary of State shall have *special regard to the desirability of preserving* the building or its setting or any features of special architectural or historic interest which it possesses.'

51 English Heritage/Countryside Agency, *Living Buildings in a Living Landscape: An English Heritage and Countryside Agency statement on traditional farm buildings* (English Heritage/Countryside Agency, Cheltenham, 2006). This replaces English Heritage's guidance of 1993 (*The Conversion of Historic Farm Buildings: An English Heritage statement*) which made a strong presumption against residential use.

52 *The Conversion of Traditional Farm Buildings: A guide to good practice* (English Heritage, London, 2006). For this and other work on farmsteads outlined in this paper, see English Heritage's HELM website www.helm.org.uk/server/show/nav.9495 and www/english-heritage.org.uk/characterisation/farmsteads (accessed 07/01/2007).

53 Gaskell and Owen, *Historic Farm Buildings*, p. 93.

54 Including National Parks, AONBs, local authorities and the Country Land and Business Association.

55 The Preliminary Regional Character Documents can be viewed at www.helm.org.uk/server/show/category.10116 (accessed 07/01/2007).

56 The Hampshire Downs, for example, were an area of large and capital-intensive farms, there being extensive evidence for large barns and courtyard layouts dating from the seventeenth century and sometimes earlier. The Hampshire Downs have

some of the earliest courtyard plans found in the country, dating from the seven-teenth century, and complete examples including large barns, stabling and a granary are rare and should be retained.

57 The mapping of farmsteads took only thirty-five days. J. Lake and B. Edwards, 'Farmsteads and landscape: towards an integrated view', *Landscapes* **7.1** (2006), 1–36; J. and B. Edwards, 'New Approaches to Historic Farmsteads', *Landscape Character Network News*, Issue 22 (Spring 2006) – link to article www.landscapechar-acter.org.uk (accessed 07/01/2007).

58 Basingstoke and Dean District Council produced a supplementary planning docu-ment (*Farm Diversification and Traditional Farmsteads: Draft supplementary plan-ning document*) using the results of this work, and others are at the time of writing being developed prior to the preparation of a template for use by local authorities.

59 P. Gaskell and M. Tanner (1998) 'Landscape conservation policy and traditional farm buildings: a case study of field barns in the Yorkshire Dales National Park', *Land-scape Research* **23**, 3, 289–308.

60 P. Gaskell, *A Study of the Social and Economic Impacts and Benefits of Traditional Farm Building Repair and Re-use in the Lake District ESA*. Report by ADAS and University of Gloucestershire for English Heritage and Defra (Cheltenham, 2006). Summarised in *Heritage Counts*, pp. 64–5.

61 P. Bibby, *Land Use Change at the Urban–Rural Fringe and in the Wider Countryside: A report prepared for the Countryside Agency* (University of Sheffield, Sheffield, 2006); see also J. Shepherd and P. Bibby, 'Developing a new classification of urban and rural areas – the methodology' www.e-consultation.net and www.statistics.gov.uk/geography/nrudp.asp (accessed 07/01/2007).

62 The RCHME survey of houses in Kent found that areas characterised by more evenly spread wealth, which are strongly associated with landscapes of dispersed settle-ment, had the highest rates of survival of late medieval houses. S. Pearson, *The Medieval Houses of Kent: An historical analysis* (RCHME/HMSO, London, 1994), pp. 141–4.

63 Land Use Consultants, *Planning for Sustainable Settlements in the High Weald AONB: Evidence-based policy-making*. Report for High Weald AONB Unit (Bristol, 2005). Smaller settlements in the more remote south of the AONB have the great-est functional strengths in sustainability terms, and the greatest numbers of resi-dents combining working from home with occasional trips out of the area. The high value of historic farmsteads and other properties and the protected status of such areas contribute to the fact that rural communities are having a less mixed or bal-anced range of buildings and households.

64 J. van der Vaart, *Boerderijen en platteland in verandering* (Fryske Akademy, Leeuwarden, 1999). My thanks to Jacob for discussion over the findings of this report.

Appendix: An assessment framework for farmstead buildings

The key options for any building comprise:

1. no action
2. dereliction/abandonment
3. demolition and salvage of materials
4. continued maintenance for structural integrity
5. full repair/restoration
6. adaptive reuse – to either agriculture, commercial, community (including recreational or educational use) or domestic purposes.

These options are both constrained and enabled to different degrees by the sensitivity to and capacity for change of the landscape setting, the farmstead and individual buildings, which are in turn a direct product of their inherent character and broader locational context.

Practical factors

These should be considered at the initial stage, as factors such as condition and access to highways and services can directly inform the range of options available.

Character

The character of a farmstead is the result, firstly, of the date, function and form of its component parts and structures, the wider layout of the steading and the context providing its relationship to the wider landscape. It is useful to identify the most dominant visual characteristics, working from landscape towards the steading as a whole and finally to the size, range and date of the buildings within the group. Less prominent or more complex features, namely detailing (internal and external) and any phases of development (for example, where an earlier timber frame hides within a stone or brick skin), can then be identified.

Sensitivity to change

Different farmsteads, buildings and their associated landscapes will have different *inherent* capacities for change, which constrain or enable them to change in different ways. The information from the character appraisal can directly inform consideration of which elements, because of their prominence, ability to adapt, contribution to local character and regional and national context, are sensitive to change, and to what degree. Some buildings, or parts of buildings with significant interior fabric or fittings, will have little or no capacity for adaptive reuse, on account of their scale (such as pigsties or dovecotes) or location. They may, however, form part of a group where other buildings have potential for adaptive reuse.

The determination of issues will benefit from a clear understanding of local circumstances, and in particular local plan policy and economic factors. Issues relating to buildings can then be set out in a clear and transparent manner, identifying any areas of conflict, and explaining how alternatives have been explored and rejected and how they have informed the approaches taken to individual sites. These can be very rapidly determined, but if adaptive reuse or restoration was identified as an option, then one could examine the issues in greater depth and draw upon a more detailed assessment framework and published guidance.

Having identified the most suitable and sustainable options, matters to be taken forward for discussion and consideration can be identified. For example:

- the salvage of materials if dereliction is an option
- the key features to retain and enhance (for example, through sympathetic fenestration or infill) if adaptive reuse is an option
- conflict with local plan policy on development outside settlement cores if adaptive reuse is identified as an option but the site is located within a landscape of dispersed settlement.

Practical Factors relating to the site
Ownership
Present Use
Condition
Scale
Location
Access: highways, services
Designation
Nature conservation/Biodiversity

Character
Landscape and Settlement
Physical form of landscape
Type and density of settlement
Historic land use

Farmstead/Building
Overall form of farmstead
External form and detail
Interior form and detail

Links to Character section of Area
Character Statement.

Value and Sensitivity
Setting and Visibility
Completeness and Coherence
Local Character and Context
Association
Local / regional / national significance

Links to Guidance section of Area
Character Statement.

Issues
The capacity for change of the farmstead and its landscape, and an understanding of
local circumstances and issues, including local plan policy and economic factors.
Links to Issues section of Area Character Statement.

Options
No action
Abandonment and Dereliction
Demolition and salvage of materials

Continued maintenance for essential structural integrity
Partial repair/restoration
Full repair/restoration
Enabling development – including sympathetic extension/new build
Adaptive reuse – in agriculture, commercial, community (including recreational or
educational use) or domestic purposes
THESE CAN BE RE-EXAMINED IN FURTHER DETAIL

Matters to take forward in discussion and negotiation

The assessment process for farmstead buildings

6 Sustainable reuse of historic industrial sites

Keith Falconer

Sustainable reuse of industrial sites is nothing new – industrial sites have a very long record of being reused for purposes entirely different from those for which they were built. Indeed, while few industrial buildings survive from the medieval period in recognisable form, the fabric of warehouses, workshops and maltings is present in a great many buildings of that time. This phenomenon is very much more evident with regard to industrial buildings of the last two centuries. For example, much of the heritage of the west of England textile industry has survived because buildings were reused for other purposes when the industry contracted a century ago. Across the whole country warehouses, maltings and mills have been reused rather than demolished purely because they offered cheap, easily utilised space. For reasons of economy there was often minimal intervention, but there was seldom any respect for the character or integrity of the building. Only in the last fifty years, coincident with the development of industrial archaeology, has there been much regard for sympathetic treatment of historic industrial buildings. It is with that evolving appreciation that this chapter is concerned.

The considered reuse of industrial buildings

In Sherban Cantacuzino's seminal book *New Uses for Old Buildings* (1975), industrial sites account for nearly half the examples and the author pays homage to work of James Richard and Eric de Mare in establishing the interest in 'functional buildings'. That those buildings were perceived to 'offer [the architect] considerably more freedom' than the conversion of other types of historic buildings was a sign of the times.

Twenty-five years later the second seminal book on this topic, *Industrial Buildings: Conservation and Regeneration*, was to be published posthumously, dedicated to Michael Stratton, its editor. This collection of essays illustrates the advances made in the philosophy of conserving and converting industrial buildings and chronicles the involvement of the Regeneration Through Heritage (RTH) initiative (see below). With its introductory essay by HRH The Prince of Wales, RTH's founder, the book demonstrated how the appreciation and profile of these buildings has risen. For anyone engaged in the reuse of a historic industrial site or the regeneration of a historic industrial district, it should be mandatory reading. As we have been so cruelly deprived of Michael Stratton's continuing contribution to this field (Michael himself would have been this chapter's appropriate author) no apology needs to be made for discussion of his own chapters, and other detailed reference to the essays in the book, to illustrate many of the points being made here.

Stratton assembled an impressive list of contributors for his book – including conservationists, international design consultants, academics, architects and entrepreneurs – and his overview essay sets the scene and illustrates how much more informed the subject had become since Cantacuzino's pioneer survey. His brief review of the decline of traditional industries, the changing attitude towards obsolete industrial sites and the attempts to provide recognition for the more significant individual sites and landscapes is masterfully summarised. After a respectful glance at the pioneer American initiatives of the 1960s and 1970s, Stratton discusses the early British attempts at reuse and redevelopment of the decayed dockland areas of our main ports, the redevelopment of the naval dockyards, airfields and barracks of the military estate, and the reuse of textile mills across the country. He then traces the gradual change in emphasis from reuse of individual sites to regeneration, with the attendant much wider and holistic canvases of interest.

The involvement in the 1980s of the dozen urban development corporations, costing around £3 billion of public money, had a huge impact on urban landscapes but, with a few honourable exceptions such the Albert Dock, Liverpool, did not produce many exemplary developments. Much more satisfying from a conservation perspective were the 357 Conservation Area Partnership Schemes (CAPS) championed by English Heritage. Levering in funds from a wide variety of sources including the Heritage Lottery Fund, the European Union and the Single Regeneration Budget, from 1994 these schemes witnessed the transformation of old harbour and warehouse areas such as Whitehaven, Cumbria and Hull while numerous smaller schemes benefited other industrial sites.

The 1990s witnessed a further development in the regeneration arsenal – the concept of genuine sustainability. Encapsulating a philosophy of low-level intervention, integration of work, leisure and urban living, and due regard for existing built-environment assets, sustainable planning seeks a balance between conservation and lasting worthwhile progress. Championed by bodies such as SAVE Britain's Heritage and epitomised by community-led developments such as that focused on the Art Deco Oxo Tower, South Bank, London, sustainable regeneration became the watchword for the development of obsolete historic industrial sites. Ironically, some of the prestigious millennium projects being developed at that time and generously funded by Lottery money paid scant regard to the precepts of sustainability and have since suffered the consequences.

By the end of the twentieth century English Partnerships, the national regeneration agency established in 1994, had emerged as the largest player involved in regeneration. It had its own investment fund and could act as its own developer, and its impact was immediately impressive. Within four years, no fewer than 2700 projects were under way, with a programme budget of £235 million and estimated to attract £630 million of private investment and create 910 000 square metres of commercial and industrial floor space. The twenty city regeneration schemes, building on the work of CAPS and the urban development corporations in cities such as Bradford, Nottingham, Bristol, Manchester and Newcastle, accounted for many of these projects. The success, for example, of the Grainger Town programme in linking the heart of Newcastle to the River Tyne was spectacular. The Newcastle waterfront, with its mix of old warehouses, Victorian commercial buildings and modern offices and apartments, has been transformed and its effects have rippled across the river to Gateshead where the BALTIC Centre for Contemporary Art is housed in the striking silo building of the former Baltic Flour Mills (Figure 6.1).

In the new century the work of English Partnerships continues across the whole regeneration spectrum. In 2004–5 English Partnerships invested over £480 million in regeneration projects and levered in an equivalent amount of private sector investment. As well as developing their own portfolio of strategic projects, they act as the government's specialist adviser on brownfield sites, they ensure that surplus public sector land is used to support wider government initiatives, and they are a key player in the Urban Renaissance programme. Much of this work is dealing with cleared sites and is targeted at providing new housing rather than the reuse of existing buildings. However, a consequence of the immense scale of many projects is that they encompass some existing industrial buildings of note such as the

Figure 6.1 BALTIC Centre for Contemporary Art, Gateshead, the former Baltic Flour Mills.

historic Market Hall building in the former Ministry of Defence (MoD) Stores Enclave in Devonport, Plymouth. The regeneration of decayed dockland areas still features strongly in the work of English Partnerships, with projects at Kings Waterfront, Liverpool, Millbay Docks, Plymouth, North Shore, Stockton-on-Tees and Humber Quays in Hull. Though much of this will be new build, historic buildings and dock structures will be preserved.

As part of its national brownfield programme English Partnerships, in conjunction with the Office of the Deputy Prime Minister (ODPM), operates the National Land Use Database which is a key tool in identifying and classifying brownfield land. The single biggest component of brownfield regeneration is the National Coalfields Programme which has a ring-fenced budget of £386.5 million. Since its beginning in December 1996 this programme has bought back into use some 1300 hectares of coalfield land, and created over 400 000 square metres of commercial floor space and over 12 000 jobs. It has also ensured the survival of historic pithead buildings, as at Pleasley Colliery, Derbyshire. English Partnerships is also involved with other government bodies in the development of the Register of Surplus Public Sector Land. This sophisticated interactive database now holds information on more than 700 parcels of land totalling some 3600 hectares originating from bodies as diverse as the MoD, British Railways Board, Coal Authority and Highways Agency, and seeks to ensure that surplus assets are used as effectively as possible. The MoD, for example, through its Defence Estates Agency, has a significant disposal programme involving many historic industrial buildings.

English Partnerships works very closely with the new Regional Development Agencies (RDAs) and they are having an equally dramatic effect. For example, English Heritage's recent acquisition of the Ditherington Flax Mills in Shrewsbury, Shropshire, the first iron-framed fireproof textile mill in the world, was facilitated by funds from Advantage West Midlands, the local RDA. English Heritage can now act as a sympathetic developer, seeking a sustainable reuse which respects the character and physical integrity of these Grade I buildings. The role of the RDAs across the country is becoming ever more prominent in providing financial backing to underpin the reuse and regeneration of a great many industrial sites, some of which are of prime historic industrial interest.

A second player to emerge in the 1990s was Regeneration Through Heritage (RTH), an initiative of the Prince of Wales and initially linked to Business in the Community, the organisation supported by Britain's largest companies to promote private sector involvement in social and economic regeneration. One of the wide-ranging activities of the tiny RTH team was the creation of a database comprising a gazetteer of industrial buildings, some restored and some awaiting conversion. The database was intended primarily as a resource for specialists, community groups and local partnerships seeking new uses for industrial buildings, and contained details of over 200 projects, some of which had direct input from RTH or its sister body, the Prince of Wales's Phoenix Trust. Though the database is no longer being maintained by Business in the Community, it provided at the time a much-appreciated source of inspiration, an information resource and a pool of valuable experience. Drawing on that experience Fred Taggart, Director of RTH, contributed a forthright chapter in Stratton's book outlining in some detail the various steps involved in undertaking a regeneration project. He stressed the importance of establishing a steering group representative of all the key interests and equipped with the necessary business, legal and financial expertise as well as conservation and building skills.

Regeneration Through Heritage is now an initiative within The Prince's Foundation for the Built Environment and is co-funded by English Heritage. It continues to promote the reuse of heritage industrial buildings at risk. Its small team, supported by voluntary experts, can travel anywhere in the UK and assist community partnerships in developing project proposals for particular buildings on the basis that the regeneration will have a catalytic effect on the wider area and benefit the local community. Projects made possible by RTH can claim to have rescued seven major historic industrial buildings, attracted inward investment of £32 million, reused 50 000 square metres of floor space and created 1100 jobs, all for the outlay of only £550 000 of public funds to meet its core operating costs. RTH's successful projects include the conversion of the derelict Harvey's Foundry at Hayle, Cornwall, into offices, craft workshops, community facilities, a heritage centre and a backpackers' hostel; the conversion at Sowerby Bridge, West Yorkshire, of canal warehouses into workshops, offices and a café and heritage information centre; the ambitious redevelopment of the vast complex of Quayside Maltings at Mistley, Essex, for apartments, a restaurant, a pub, offices, workshops and a managed performance and meeting space; and the reuse of Houldsworth Mills, the impressive Grade II* cotton mills in Stockport, Cheshire, to provide offices, a managed workspace, a college campus, a health club, a nursery, community space and shared-ownership apartments.

When Stratton's book was published in 2000 the debate about the role of statutory designation in managing change was in its infancy. Battle lines were still drawn between purist conservationists and radical developers, with English Heritage caught in the middle. Five years into the new millennium English Heritage spurned its perceived image as the 'conservation police' and fully embraced the concept of the sympathetic management of change (see also Chapter 3). Encouraged by the enthusiastically warm reception of the sector's *Power of Place* (2000) and the equally heartening response from government, *The Historic Environment: A force for our future* (2001), English Heritage, with the support of the Department for Culture, Media and Sport, is pioneering a new designation regime based on a single register of historical assets. Fundamental to this regime are the concepts of mutual management agreements and enabling sustainable developments. This creates substantial opportunities for the regeneration of large historic industrial sites, especially those of the twentieth century where strict statutory planning controls might not be appropriate.

With this enlightened regeneration context firmly established in the planning agenda, the implications for heritage industrial buildings are promising. The lead given by *Industrial Buildings – Conservation and Regeneration*, with its review of the factors concerned in regeneration, the methodologies employed to achieve successful projects and the identification of good and bad practice, is worth following. With many strong-minded authors, the book tends to be prescriptive, even at times formulaic, and some points are made repeatedly but usually from the different perspectives fashioned from hard-won experience. It is, however, worth sticking with the anecdotal case studies assembled from all over the world as many of them are extremely instructive.

Stratton himself sets the tone in his second chapter where he seeks to 'Understand the Potential' by examining the location, configuration and conversion options. He outlines the particular problems and opportunities posed by the location of industrial sites in various contexts – rural, urban and suburban – and demonstrates how even problems of location can be turned to advantage. The discussion of the constraints and potential of various industrial building types – multi-storey mills, single-storey weaving sheds, warehouses, maltings, breweries, railway stations, railway engineering works, twentieth-century day-lit factories, aircraft hangars and steelworks – shows that there is scarcely a building type that has not been subject to conversion, with varying degrees of success and intervention. Textile mill complexes, for example, with their mix of building types, have witnessed a wide spectrum of reuse. This can range from relatively slight intervention as at Saltaire, near Shipley, Bradford, West Yorkshire (Figure 6.2), where the open floors of the main mills are used for retail, exhibition and catering space and the top-lit weaving sheds are most appropriately used for assembling electronic components, to the drastic intervention as at Ebley Mill, Stroud, Gloucestershire, in the conversion to local authority

Figure 6.2 Saltaire, near Shipley, Bradford, West Yorkshire.

Figure 6.3 Ebley Mill, Stroud, Gloucestershire, converted to local authority offices.

offices (Figure 6.3). Even unpromising subjects such as steelworks have been subject to regeneration schemes of varying intensity internationally, ranging from the National Historic Landmark treatment of the Sloss Furnaces in Birmingham, Alabama, USA, through the 'public venue' landscaping of the Völklingen and Emscher Park steelworks in the Ruhr area of Germany, to the Magna 'discovery centre' conversion of the Templeborough Melting Shop at Rotherham, South Yorkshire, once the largest steel-making plant in Europe.

In his discussion of approaches to conservation and future uses, Stratton does not pull any punches. He is rightly critical of the breed of consultant that proposes ludicrous combinations of far-fetched uses for almost any derelict site designed to attract European and Lottery capital funding based on inflated visitor projections, and he argues for a return to basics. The need for new uses to be financially sustainable had been demonstrated in the 1980s by analyses such as Eley and Worthington (1984) and in practical advice from URBED (Urban and Economic Development Group) (1987), a not-for-profit urban regeneration consultancy, but had been too often ignored. Both Nicholas Falk of URBED and John Worthington of DEGW plc Architects & Consultants were to reiterate this advice to good effect in later chapters in *Industrial Buildings – Conservation and Regeneration*. As founder director of URBED, Nicholas Falk can call on more than twenty-five years of practical experience in projects for the reuse of redundant industrial buildings. URBED's research on the subject has embraced 'over 600 examples, involving every imaginable kind of building and use, suggesting that the physical problems are capable of resolution in most contexts' (Stratton, 2000, p. 95). Therefore when Falk, and Worthington, from a different standpoint in a later chapter, both argue for the virtues of incremental development as opposed to 'big bang' solutions, the advice is worth heeding. Major regeneration schemes such as that undertaken by DEGW for the centre of the town of Jena, in former East Germany, may be achievable with the will and financial backing of the German Federal Government but work less well in the less extreme situations in British cities (Stratton, 2000, pp. 147–56).

The virtues of flexibility, realistic financial aspirations, close budget control and, on occasion, rank opportunism, are convincingly extolled by the entrepreneur Bennie Gray, founder of the SPACE Organisation (Society for the Promotion of Artistic Enterprise) (Stratton, pp. 103–16). SPACE does not just design and develop but actually runs numerous projects reusing industrial buildings, including the much-lauded Custard Factory at Digbeth, Birmingham, the Big Peg in Birmingham's Jewellery Quarter, and Canalot, Danceworks and Alfie's Antique Market in London. Though largely having to find backing from financial institutions and the occasional small grant from public funds, it has provided workspace for about a thousand small start-up companies at a fraction of the cost of many government-backed schemes. Gray's recipe for success sets out seven steps – the 'Big Idea', finding the money, getting the permissions, doing the design, construction, marketing and finally management. He demonstrates these by looking at the trials and tribulations encountered, but also at the pleasure and the satisfaction that the various projects have brought. His ad hoc and personal approach is in the same vein as Ernest Hall and Jonathan Silver's early development of Dean Clough Mills, Halifax, West Yorkshire, and is a salutary counter to some of the more ponderous and bureaucratic initiatives elsewhere.

Stratton (2000) concludes with a case-study section of sixty-six entries illustrating more than a hundred sites drawn from the Regeneration Through Heritage database. They range from single relatively small buildings such as the Bluebird Garage in Chelsea, London, to the huge complexes of the Albert Dock in Liverpool and Dean Clough and Saltaire Mills in Yorkshire. They embrace such different building types as pottery kilns and a bone-grinding mill in Stoke-on-Trent, workshops in the Birmingham Jewellery Quarter and the engineering sheds of Swindon Railway Works. A brief review of a few of these, together with some post-2000 examples of conversion, illustrates the circumstances, evolution, detail and success of different types of conversion.

The **Albert Dock**, Liverpool (Figure 6.4), is an obvious starting point as its regeneration, which has spanned a quarter of a century, encapsulates so great many of the factors affecting sustainable reuse. Jesse Hartley's Albert Dock and warehouses were constructed in 1846–8; they are the finest expression of the closed wet dock systems pioneered in Liverpool and comprise the largest group of Grade I listed industrial buildings in the country. When Liverpool's South Docks were closed to shipping in 1972, Albert Dock and its neighbouring docks were allowed to silt up and the future for the abandoned warehouses was bleak. In the 1970s the fortunes of the city of Liverpool were at a low ebb and economic unrest culminated

Figure 6.4 Jesse Hartley's Albert Dock, Liverpool, constructed 1846–8.

in 1981 in the Toxteth riots. These riots focused government attention on Liverpool's waterfront, and the Merseyside Development Corporation (MDC) was created to stimulate regeneration. One of the Corporation's first initiatives was to back a detailed survey of the South Docks so that the resource was fully understood; this led to the publication of the book *Liverpool's Historic Waterfront*, which firmly established the dock system as of supreme international interest (Ritchie-Noakes, 1985). Albert Dock had already been the object of several drastic and unsympathetic redevelopment proposals but now with central government funds a measured approach could be adopted. A masterplan devised by Holford Associates for the MDC, with the developers, Arrowcroft, recognised that complete restoration and reuse of all the enormous warehouse stacks would take many years but that confidence could be generated by rehabilitation of the structural envelopes and prestigious reuse of some elements. Thus part of one warehouse stack was converted by James Stirling, Michael Wilford and Associates into Tate Liverpool while another block was conserved as Merseyside Maritime Museum. The Dock Traffic Office became Granada Television's News Centre and some bars and restaurants opened at quayside level. This pump-priming and cultural adventure has been spectacularly successful. Twenty years later the stacks are fully occupied by apartments, offices, two hotels, tourist attractions, numerous restaurants and bars; neighbouring warehouses such as those at Wapping Dock have been converted into apartments and a marina has developed in docks to the south.

The regeneration of **Swindon Railway Works**, another key historic industrial site, followed a very different path – speculative, piecemeal and opportunist – but with similar positive results. Started by Isambard Kingdom Brunel in 1842, the Locomotive and Carriage Works of the Great Western Railway (GWR) at Swindon were, by the early twentieth century, one of the largest such works in the world and were the raison d'être of the modern town of Swindon. In the post-war years, however, the works declined under railway nationalisation and were eventually closed in 1986. The site was bought by Tarmac, a commercial developer, most of the unlisted buildings were demolished, and those workshops and offices that were listed were left abandoned. Most of the cleared areas of the site were contaminated and required extensive remedial works, and by the time that an ambitious scheme for redevelopment had been devised the property crash at the end of the 1980s rendered it uneconomic. The furthest cleared areas were then sold off for supermarket and residential development but the historic core site, apart from an office occupied by the developers themselves, remained unused. However, the Grade II listed GWR general offices in one corner of the site had been brought to the attention of the Royal Commission on the Historical Monuments of England (RCHME). The Royal Commission was seeking accommodation to centralise its activities in southern England, with space to build an archive store and close to a main line station. The GWR office building was known to the RCHME since some recording at the works in 1984 and, being part of a vast derelict works site, there was room to build a state-of-the-art archive store designed by DY Davies. The offices, when rehabilitated in 1994, became the RCHME's headquarters and now, with the adjoining 1842 works building, constitute English Heritage's Central Office, housing the National Monuments Record and much of English Heritage's commercial and research staff (Figure 6.5).

Figure 6.5 The GWR General Offices, Swindon, now housing English Heritage and the National Monuments Record.

The enormous Grade II* sheds of the locomotive works built by Joseph Armstrong in the 1870s remained empty until Joe Kaempfer, chief executive of the American firm McArthurGlen Designer Outlet, had the vision to convert them into an outlet shopping mall (Figure 6.6). The Great Western Designer Outlet Village, which opened in 1997, contains over a hundred units, attracts some 4 million shoppers a year and has revitalised this part of central Swindon. The quality of its conversion, insisted upon by English Heritage and the local authority and guided by *Swindon: The Legacy of a Railway Town*, has led to its being regarded as an exemplar, and its success has encouraged other developments on the site (Cattell and Falconer, 1995). The Pattern Store has been converted into a bar and restaurant, STEAM – Museum of the Great Western Railway occupies another historic workshop, and the National Trust has built its headquarters alongside. (A full account of the circumstances of this redevelopment is contained in Falconer (2000).)

At the same time as these large-area regeneration schemes were being undertaken by public agencies and major development firms, there were equally influential single-site conversions which were the result of individual entrepreneurs or specialist firms. As we have seen, *Industrial Buildings – Conservation and Regeneration* details the successful initiatives of Ernest Hall and Jonathan Silver at Dean Clough, Silver himself at Saltaire and Bennie Gray at various sites in London and Birmingham. More recent spectacular conversion projects have been undertaken by a specialist firm that emerged in the late 1990s and has now become synonymous with adventurous apartment developments – Urban Splash.

Figure 6.6 The sheds of the locomotive works, Swindon, converted into an outlet shopping mall.

Figure 6.7 Lister's Manningham Mills, Bradford, Yorkshire.

Urban Splash was set up in 1993 by Tom Bloxham and Jonathan Falkingham and they made their name initially by their innovative projects in Manchester and Liverpool. In Manchester their early work at Smithfield Buildings led to the conversion of Box Works and Albert Mill in Castlefield and Ducie House and Waulk Mill at Ancoats. In Liverpool the success of their Concert Square project in 1993 led to much-praised conversions of the Matchworks in the run-down Speke district and the Tea Factory, the Vanilla Factory and the Liverpool Palace in the Ropewalks area of the city. The company has now blossomed in other parts of the country, and this chapter will conclude by describing three of Urban Splash's projects.

Lister's Manningham Mills in Bradford, Yorkshire (Figure 6.7), in the words of the City Council, 'had become a symbol of decline and of the perception that we could not deal with regeneration . . . Urban Splash has provided hope which will unlock the potential.' The striking mill complex, with its Italianate chimney and associated park and Cartwright Hall Art Gallery, crowns a hill above Bradford and was once the largest silk mill in the world, employing 11 000 workers. Derelict and decaying for over twenty years, it was one of English Heritage's most problematical Grade II* 'Buildings at Risk', and EH allocated 420 000 euros towards the cost of essential repairs, Yorkshire Forward, the RDA, 5.6 million euros, and Bradford City Council 2.8 million euros. The new uses include commercial offices, studios, community space and leisure activities as well as apartments. The first phase of 130 luxury apartments was reserved within days of coming on the market. They are creating a new vibrancy in an area much in need of regeneration.

Fort Dunlop (Figure 6.8) is situated alongside the motorway on the main approach to Birmingham, and as the *Birmingham Post* recorded, 'First impressions of England's second city are about to change – thanks to Fort

Figure 6.8 Fort Dunlop, Birmingham.

Figure 6.9 Sir John Rennie's grand naval King William Yard, Plymouth, built in the early 1830s.

Dunlop, Birmingham's latest landmark.' The 70 million euro conversion of the huge derelict storehouse includes offices, retail and a hotel. Urban Splash acted as developers for the Regional Development Agency, Advantage West Midlands.

King William Yard, Plymouth (Figure 6.9), was a grand naval victualling yard built by Sir John Rennie in the early 1830s and is one of the largest

Grade I listed sites in the country; the ten buildings total over 48 000 square metres. It was acquired in 1999 by the South West Regional Development Agency, and the agency went into partnership with Urban Splash to develop this problematical site which is surrounded on three sides by water. English Partnerships contributed £4.25 million towards repairs on the Mills and Bakery buildings. A restaurant, wine bar and arts centre form part of the scheme on the ground floor of the Brewery and all the luxury apartments in the first phase – the Brewery and the Clarence Range – were again sold within days.

The marketing of all these projects and the company itself is impressive. Urban Splash have produced publicity material for these last three projects, each with a CD-ROM outlining the past achievements of the company and extolling the attractions of the adventurous conversions. From the standpoint of a movement that has been lobbying for recognition of the potential of historic industrial sites for many years, such packaging is gratifying.

As these case studies illustrate, the options for reuse are legion. The commonest and often most interventionist are apartment conversions, followed by office conversions, but in the hands of high-quality designers these too can be sympathetic and certainly exciting. A mixture of uses, however, is increasingly recognised as the most appropriate solution for large industrial complexes, while an incremental rather than 'Big Bang' approach is now seen to be more effective.

References and further reading

Cantacuzino, Sherban, *New Uses for Old Building* (Architectural Press, London, 1975).

Cattell, John and Falconer, Keith, *Swindon: The Legacy of a Railway Town* (HMSO, London, 1995).

Eley, Peter and Worthington, John, *Industrial Rehabilitation* (Architectural Press, London, 1984).

Falconer, Keith, 'Swindon's head of steam: the regeneration of the GWR's works', *Patrimoine de L'industrie* 3, K. (2000), 21–8.

Glassberg, David, 'Presenting memory to the public: the study of memory and the uses of the past', *Cultural Resource Management* 21, no.11 (1998), 7.

Ritchie-Noakes, N., *Liverpool's Historic Waterfront* (HMSO, London, 1985).

Stratton, M. (ed.), *Industrial Buildings: Conservation and regeneration* (E & FN Spon, London, 2000).

URBED, *Reusing Redundant Buildings: Good practice in urban regeneration* (HMSO, London, 1987).

7 Realms of memory: changing perceptions of the country house

Giles Waterfield

The title for my chapter derives from the seven-volume *Lieux de mémoires* by Pierre Nora, translated into English as *Realms of Memory*. Published between 1984 and 1992, Nora's compilation of essays by numerous scholars explores the 'memory places of French national identity as they have been constructed since the middle ages'. In contrast to history, which is seen as a more intellectual practice based on the study of empirical reality, the collective memory is interpreted by Nora as the study of 'national feeling not in the traditional thematic or chronological manner but instead by analyzing the places in which the collective heritage of France was crystallized, the principal sites . . . in which collective memory was rooted'. The book examines such abiding themes as the Gauls, the cathedral, street names, the Vichy government. Nora's book has inspired similar compilations in the Netherlands, Germany, Denmark and Italy among other countries – though not, at least not in quite the same form, in either the United States or Britain.

If a series called Realms of Memory were to be published about the British Isles – or more manageably perhaps about England since Scotland and Ireland have such different collective memories – one could be relatively sure that country houses would form one of the sections. The story of the changing collective view of these places has been addressed by Peter Mandler in *The Fall and Rise of the Stately Home*, published in 1997. Mandler shows how in the twentieth century the country house came to be seen as an important element in the national heritage, possibly, as he phrases it, 'the country's greatest contribution to Western civilisation'. He studies the cult of Merrie England associated with medieval and Tudor buildings in the early nineteenth century through literature and the visual arts, the exclusion of Palladian architecture from this approval until well into the twentieth century, the ambiguous attitude to great aristocratic houses expressed in many circles at least until the post-war period, and the creation of the notion of country houses as a crucial aesthetic achievement from the 1960s onwards. This chapter aims, briefly, to suggest how collective memories of the country house developed during the twentieth century. It considers how far the battle for the preservation of the country house has been won, not so much in practical terms as in popular and official perception, and asks whether the enthusiasm and respect for the country house that emerged in England in the 1970s have been firmly consolidated, or whether, as Mandler suggests, the country house cult represented a 'fashion'.

The chapter is written from the point of view of a believer in, and enthusiast for, the value of preserving country houses, not only as buildings but as entities. But the writer sometimes strays outside the temple, to places where the transcendent value of these houses, their collections and parks, is questioned – or perhaps left aside, omitted from discussion, as though the participants in debates around heritage were unwilling to engage in such questions (a trend one can sometimes observe in funding bodies such as the Heritage Lottery Fund). More vigorous scepticism is also evident, as in an essay published in 2006 in *The Uses of Heritage* (edited by Laurajane Smith). In this essay, entitled 'Knowing your place: landscapes of class, deference and resistance' – not a title that suggests warm feelings towards the historic house – Laurajane Smith and Gary Campbell analyse parallel surveys of visitors to country houses and to trade union museums. Many of the findings, about the types of visitor in terms of age, for example, were relatively predictable: the highest proportion of visitors were what is known as 'silvers' (a reference to hair rather than finances). What was less predictable were the statements about what people found most pleasurable in their country house (as opposed to museum) experience. A majority of the country house public – over 60% – said that what they most appreciated was the sense of comfort that historic houses provided, otherwise interpreted as a feeling of reassurance, escape, a return to older values, peacefulness. The questioners were apparently not impressed by this attitude on the part of the 'silvers'. They were still less impressed by a group of ladies observed at Harewood (ironically, a house with an outstanding record in the field of learning and interpretation) who reacted negatively to the arrival in the state rooms of a group of schoolchildren from Leeds. The reported reaction of these ladies was to ask what the children were doing there; the place was not for them. In conclusion, the essay rejects what are seen as the elitist and backward-looking messages emanating from the country house.

Leaving these comments aside for a moment, I would like to outline some of the changing attitudes to the country house that our imagined essay in the British Realms of Memory might delineate. In the late nineteenth and early twentieth centuries, architectural value was ascribed primarily to earlier buildings, that is to say those built before around 1714. This view – official and popular – died hard (though it is odd to reflect that at the same period prices for English eighteenth-century portraits became enormously high). After all, Georgian architecture has only relatively recently been generally included in the 'country house' pantheon, probably since the 1940s. The Victorian house has been admitted since the 1980s (it is hard now to recall the doubts that the acquisition of Norman Shaw's great house at Cragside in Northumberland aroused at the National Trust in the 1980s). Nevertheless, the foundation of *Country Life* in 1897, and the writings in the early twentieth century of such figures as Avray Tipping and

Christopher Hussey (both closely associated with *Country Life*), stimulated a new academic interest in the decorative arts and in phenomena such as the Gothic Revival and the Picturesque Movement, which had to a great extent been forgotten. This academic interest in the history of houses and interiors was furthered in the middle of the century by certain seminal writings. These included John Summerson's books on Georgian architecture, notably *Georgian London* (1943), and Howard Colvin's *Dictionary of British Architects*, published in its first version in 1954. Both stimulated a spate of academic studies on British architecture. Summerson's book was originally intended to form a series of lectures at the Courtauld Institute of Art, and it is no accident that the creation of the first university institution for the study of the history of art in Britain encouraged a more rigidly academic interest in the field.

This academic interest in the subject encouraged a growing nostalgia for these houses and the way of life they represented, a complex nostalgia which was expressed in various literary forms. Many of the earlier twentieth-century writers who used country houses in their books concentrated on the great medieval and Tudor houses. Both Virginia Woolf in *Orlando* (1928) and Vita Sackville-West in *The Edwardians* (1930) celebrated Knole – though in both these cases the writer's affection for the house and its past inhabitants was flavoured by a certain irony about its recent history. This romantic literary nostalgia developed as a cult through the twentieth century and is perhaps most famously expressed in *Brideshead Revisited*, written in 1944 as an elegy for the great house and its way of life. Interestingly, Brideshead is a Palladian house: Evelyn Waugh, energetically ahead of his time in matters of aesthetic appreciation, did not restrict himself to the long-lasting Merrie England tradition.

But Waugh belonged to an elite group, and his nostalgic views were by no means widespread. After the Second World War country houses to most people seemed to be finished, at a time when there was a fairly general hostility throughout Europe to the traditions represented by the historic palace or house. This hostility was expressed at its most extreme in communist countries such as Czechoslovakia where the contents of such buildings, tainted by their association with an inimical aristocratic order, were often dismantled. It was also apparent in western Europe. At the Royal Palace in Genoa, as reinstalled for visitors around 1950, references to the Royal House of Savoy were as far as possible eliminated, with the rooms presented as a series of art galleries, containing important works of art but not habitable or significant as ceremonial spaces. This approach generally still applies in Italy where it is usually very difficult to gain any understanding, in terms of quotidian presentation, of the original function of the rooms within a historic palace that has been converted into a gallery. A similar approach is apparent in England, at much the same period, at Apsley House: when the house was taken into public ownership and prepared for regular public visiting after the war, it was given, as far as possible, a non-domestic character. Art galleries were considered to be socially neutral spaces (in a way that would probably no longer apply today) whereas noble houses were not.

Even though such an extreme line was not taken in newly acquired National Trust houses by James Lees-Milne and his colleagues from the 1940s onwards, attitudes there were not altogether different. The rooms were to be shown as though the owner still lived there and had just gone out for the afternoon. More revealingly, they were to be purged of later accretions, and often furnished instead with what were regarded as superior pictures and furniture brought in from other properties. There was no recognition of the social fabric and gradual development of the buildings. Equally, the history of life below stairs or in the stables was thought to be of no interest. James Lees-Milne languidly records the removal of the contents of the servants' rooms from Attingham Park in the 1950s. At the same period the entire contents of the servants' bedrooms at Saltram, bought by the Francophile Earl of Morley in France in the early nineteenth century and of extraordinary rarity, were thrown away. A similar approach applied to publications, which ignored people and looked at buildings in isolation. It was the gallery quality, the existence of the works of art, that counted: in his Buildings of England series Nikolaus Pevsner often gave the impression, through his minimal or non-existent account of the families that had built historic houses, that they had never been inhabited.

At the same time country houses came to be embraced by popular culture. In the 1950s, many commentators in the press, and the public in general, were fascinated by what appeared to be a new entrepreneurial spirit of aristocratic owners, whipped up by such dashing figures as the Marquis of Bath at Longleat, and the Duke of Bedford at Woburn Abbey, who recorded his salvation/degradation of the house in his autobiographical work *A Silver-Plated Spoon*. This fascination extended to academic writers. Here is Pevsner writing, with a faintly repelled relish, about Woburn Abbey in his *Bedfordshire* volume, in 1968:

> ... ever since the Duke of Bedford opened Woburn Abbey to the public in 1955, Woburn has become the case par excellence of mass attraction. In 1956 c. 475,000 visitors were counted, and the side-shows included a zoo with e.g. bison and many species of deer, a pets' corner, model soldiers, and sailing. It goes without saying that the majority of the visitors care more for the entertainments (including a glimpse of the Duke) than for the house ...

(It may be worth noting that in the revised edition of *Wiltshire* in 1975, when such activities were more familiar, Nikolaus Pevsner's account of Longleat, revised by Bridget Cherry, makes no mention of twentieth-century commercialisation other than to comment on 'the sub-Freudian phantasmagoria' of Lord Weymouth's murals.) Pevsner's uncharacteristic piece of contemporary commentary (as well as his lack of interest in what today would be hailed as access and inclusiveness) reflects a widespread perception of historic houses and their owners at the time. Though the perception of overt commercialisation may have gone today, some of the attitudes that were engendered – not so much the grudging admiration for showmanship as a sense that country houses owners were wily manipulators of the public purse – have persisted.

These confusing trends initiated what might be described as the Golden Age of the Country House, beginning around 1970. The landmarks are familiar, notably the energetic campaign for the recognition and preservation of historic houses waged by such heroic figures as John Harris and John Cornforth, which led to the 1974 exhibition The Destruction of the Country House at the Victoria and Albert Museum, as well as to a range of publications. The museum was also responsible for an active research programme which raised understanding of the actual rather than the mythic character of historic interiors. The pioneering work of the Furniture and Woodwork Department of the Victoria and Albert Museum, under the keepership of Peter Thornton from 1966 to 1984, was responsible for a vigorous rethinking of the interiors of historic houses, put into practice in the revolutionary refurbishment of Ham House. This interest was also stimulated by important research in the field of social history, triggered by Mark Girouard's *Life in the English Country House* (1978).

These various activities stimulated the rise of the country house nostalgia industry. This took various forms. The Treasure Houses of Britain exhibition at the National Gallery of Art in Washington D.C., in 1985, was much criticised as providing a shop window for historic artefacts, but stimulated further investigation of the background to historic buildings. On a more playful level, the television film of *Brideshead Revisited* made a major impact. Changing attitudes to social history stimulated a new approach to display. This included presenting, to an enthralled public, the servants' quarters and the social organisation of a great house, notably at Erddig in Wales, which opened to visitors in 1972. Original research, resulting in publications which could appeal to a specialist and a general audience, was a crucial element in these innovatory activities – the importance of this component in the generation of interest in the country house tends to be forgotten.

As a result of these changes in perception, and particularly the growing realisation of the value of country houses as entities, by the 1980s the country house, and particularly the Georgian house, had indeed reached the status of national icon that had previously been denied it. Though the National Heritage Memorial Fund defined 'heritage' very broadly, the purchases for the National Trust (acting as representative of the nation) of Kedleston Hall, Weston Park and Calke Abbey in the mid-1980s, with very substantial support from the Fund, were subjected to little or no public questioning as they might have been twenty years earlier or later. Country houses, it was generally agreed, were the best. It has been suggested that at a time when there was much popular interest in property, and particularly in the ownership of historic property, the idea of the great or small country house, with its original furnishings, appealed to many individuals.

So what is the situation now – with regard not to the future of the collections but to the collective view and memory of these sites? What place do country houses hold in the affections of the public, both the official public and the general public? Can we be confident that they have achieved the secure and long-established status of, for example, cathedrals? Are they even as safe in public esteem as examples of industrial archaeology which in the past forty or so years have gained such a strong hold on our consciousness, whether settlements such as Ironbridge or historic canals? If another Kedleston came along, could believers in the importance of the country house be confident that government or funding bodies would react as they did to Tyntesfield? Government, after all, has been shielded by the Heritage Lottery Fund from playing any active part, restricting its support to the mouthing of benevolent platitudes about access.

The current gradual but steady decline in the number of visitors to country houses (rather than gardens or special events) which is tracked by the Historic Houses Association is not the only indication that the country house is out of fashion. A lessening of interest is also evident in terms of art historical research, as suggested by applications for research grants to the Paul Mellon Centre for Studies in British Art, which at present seldom focus on British country house architecture and design. Many of the most important works in this field in recent years have tended to concentrate on economic history, as in Richard Wilson's *Creating Paradise: The building of the English country house*, published in 2000, or on aspects of the estate, such as Giles Worsley's *The British Stable* (2004). At a more popular level, it is not insignificant that at auction the market for 'brown furniture' has dropped considerably in the past five years (by about 40%, it has been estimated, for good-quality pieces outside the stellar range). Traditional English stuff is just not fashionable. Although there may be no hostility to the past involved in this phenomenon, the loss of interest in this particular version of England reflects the speed, and unpredictability, with which tastes change.

What is more, although the aristocracy can hardly any longer be seen as offering any form of political threat, suspicion of these houses lingers on, fuelled by the fact that so many country houses, and particularly the really large ones, remain in private hands or in the hands of family trusts. The view that the preservation of landed estates for heritage reasons actually represents a covert defence of privilege and inherited wealth has been around a long time. Equally, the idea that country houses perpetuate an unhealthy obsession with an artificial concept of heritage, providing a means by which this country has avoided coming to terms with the real patterns of change and decline, has repeatedly surfaced in the work of such cultural critics as David Lowenthal and Robert Hewison. As Hewison wrote in *The Heritage Industry* (1987): 'The National Trust's commitment to the continued occupation of houses for whom it accepts responsibility by the families that formerly owned them has preserved a set of social values as well as dining chairs and family portraits.' A more aggressive point of view was put by the historian and journalist Tristram Hunt, in a review of what he saw as the aristocracy's current exploitative mania for selling major works of art. In the Winter 2005 edition of *Quarterly*, the Art Fund's magazine, Hunt writes that after one band of noblemen had fallaciously established themselves as 'the formal depository of our national heritage . . . [today] another generation of aristocrats are exploiting their positions and playing the market to cash in on our cultural heritage . . . they are showing themselves in a different light: as large, landowning corporations focused on instant shareholder value'.

But there are some positive aspects here. In the past decade the concept of 'heritage' has expanded, not least under the influence of the National Heritage Memorial Fund (NHMF), founded in 1980, and its sister organisation, the Heritage Lottery Fund, which recently celebrated its tenth anniversary. The definition of heritage espoused by the NHMF in the 1980s was focused on the built and the natural environment. As has been mentioned, the fund was greatly involved in the preservation of stately homes, even running the risk at times of being perceived as a body dedicated to this purpose. This approach has not been altogether abandoned, as the donation in 2002 of the largest ever grant from the NHMF for the purchase of Tyntesfield made clear. But traditional views of heritage have now been greatly enlarged, to embrace non-tactile heritage – such things as dialect, oral memories, dance, the traditions of immigrant communities.

While this expansion does not exclude the old definitions, it does mean that traditional beneficiaries of public largesse are more critically assessed. The support by the Heritage Lottery Fund for the endowment of Tyntesfield depends on the fulfilment of certain conditions including a radically new approach to display and interpretation. It will be very interesting to study, over the years, the impact of these demands by the Heritage Lottery Fund on this and other historic houses. At Tyntesfield there are two important elements in the new approach: firstly, the involvement of local communities, old and young, virtuous and not so virtuous, in planning the future of displays inside and outside the house; and secondly, a policy for carrying out the conservation of the house over a number of years, in full view of the public. These activities suggest new ideas of what the country house means – as a centre for community activities, as a site for visible conservation, and as a means of stimulating urban or rural regeneration.

Under the pressure of changing definitions of heritage, organisations and individual owners have reacted positively. As the research that the Attingham Trust carried out for the report *Opening Doors* (edited by Giles Waterfield, 2004) underlined, the role of education departments, or of individuals concerned with education, in historic buildings has been transformed in the past ten or fifteen years. The size and importance of education departments in national organisations in England and Scotland has increased greatly. In large private houses there is a new emphasis on education, notably at such places as Chatsworth and Burghley. The Historic Houses Association, albeit with modest resources, has launched a programme to encourage owners of smaller houses to engage in educational work, even though some owners find this expensive and difficult. The range of issues addressed in these houses, particularly for children, has expanded. Such questions as slavery – previously taboo – are frequently discussed: at Harewood the hidden history of the house has been opened up, literally in the case of boxes of archives containing previously unknown information about the financial dealings of the Lascelles family. At another level, the academic potential of these houses is being explored through such initiatives as the Yorkshire Country Houses Partnership. This partnership between seven great houses and the University of York, having investigated the theme of women in country houses, is now exploring the wealth of archival and material information regarding the estates and libraries of seven houses. It is hard not to feel that the strictures regarding the limitations of country house interpretation expressed in *The Uses of Heritage* overlook some of these achievements.

It does seem that the country house is gaining a new role, as a place for learning and research. There are many possibilities here though, as *Opening Doors* suggested, there are also many obstacles: shortages of funding, a reliance on primary school provision at the expense of other age groups, a lack of coordination, and a disjunction between the academic community and those who provide for the public. The recent decision by the Department for Culture, Media and Sport and the Department for Education and Skills to combine on the production of a website is a modest step forward, but it looks as though successful initiatives will continue to be the responsibility of individuals rather than any coordinating body.

Country houses have for a long time been performative spaces, spaces dedicated to enhancing and providing a backdrop to performances – whether displays of wealth or hunting prowess, royal visits, electioneering, open-handed hospitality, or just plain parties. This character continues with the part played by houses in costume drama and film – most people who recognise Lyme Park probably regard it as the residence not of the Leghs but of Mr Darcy. Country houses are constantly changing their role, and as a result public perception of them is always changing too. For those concerned with their healthy future it is vital to avoid complacency and to remember how variously they have been viewed and continue to be viewed, if the historic house is to be positively perceived as a realm of memory in the twenty-first century.

Further reading

Giles Waterfield (ed.), *Opening Doors: Learning in the historic built environment* (2004).

Duke of Bedford, John, *A Silver-Plated Spoon* (Attingham Trust, London, 1959).

Hewison, Robert, *The Heritage Industry: Britain in a climate of decline* (Methuen, London, 1987).

Lowenthal, David, *The Past is a Foreign Country* (Cambridge University Press, Cambridge, 1985).

Mandler, Peter, *The Fall and Rise of the Stately Home* (Yale University Press, New Haven and London, 1997).

National Heritage Memorial Fund, *Treasures for the Nation* (British Museum, London, 1988).

Nora, Pierre, *Realms of Memory: Rethinking the French past*, trans. Arthur Goldhammer, 3 vols (Columbia University Press, New York, Chichester, 1996, 1997, 1998).

Samuel, Raphael, *Theatres of Memory* (Verso, London, 2004).

Smith, Laurajane, *The Uses of Heritage* (Routledge, London, 2006).

Wright, Patrick, *On Living in an Old Country* (Verso, London, 1985).

8 Conserving buildings of the Modern Movement

John Winter

Around 1930 we see the first British buildings designed under the influence of continental Modernism. Typically, they tended to be cubist in form, with flat roofs and built of concrete, or designed to look as if they were built of concrete. These buildings did not weather well, and have left us with a host of technical problems. There are also philosophical problems, for they were designed with no thought for their conservation; they were designed to be revolutionary yet, only a couple of generations later, two of the houses belong to the National Trust. The poor weathering characteristics of these buildings soon became evident, and by 1938 the leading protagonists of this new architecture were already exploring more traditional ways of building, so the high noon of the movement was very brief. The buildings were mostly upper-middle-class houses. The best were included in F.R.S. Yorke's seminal book *The Modern House in England*.[1] The architects' hope that the Modern Movement would catch on and become the normal way of building did not happen, so there are relatively few – certainly fewer than a hundred – houses that need serious conservation. The Twentieth Century Society has searched hard to increase the number of examples, but previously unknown Modern Movement houses of quality almost never turn up, and one should stick to those in Yorke's book.

The houses were environmentally poor and unsuited to the British climate but, in pristine state, they are unbelievably beautiful; and they carried the seeds of the new architecture on to the next generation and so are of critical importance to our world.

Three of the houses are already open to the public.[2] That is probably enough. Another fifty are an important part of our heritage and need loving care and conservation. The rest can take their chance with the vast majority of buildings, being conserved knowledgeably when it is economically sensible to do so. Let us not be too precious about the lesser works of this part of our heritage.

Utopia

Today, 'causes' are no longer fashionable. But we do have a human need to believe in something larger than ourselves. The best houses of the Modern Movement reveal their designers' fervent belief that they were struggling for a better world; each house was seen as a step along the way.

This lost world of belief is attractive to us. It is easy to sneer at the affluent 1930s socialists designing houses with maids' bedrooms placed so as not to overlook the owners in the garden. It is also easy to ridicule a movement despising suburbia, yet whose products were mostly single-family houses with gardens. But the idealism shines through, and who can doubt the substantial sincerity of the designers when face to face with the actual buildings?

It was not only a social mission that these architects claimed. They also had a technological mission: they believed in modern materials, particularly steel and concrete. It must have been galling for them that so many of the houses could only be realised in more conventional construction, but the image of a better world was maintained by making other materials look like concrete. Because they weathered so badly, many of us first knew of these buildings in a deplorable state. There was something especially moving about the 'failed Utopia', and the buildings in their distressed state were very touching. But the buildings need to be put into good condition, and we have to accept some loss here in order that the buildings can survive for the enjoyment of future generations.

The task of conservation architects for these buildings is to ensure that they function well for today's needs, are not excessively expensive to maintain and treat the world's resources with respect, while letting the beauty and the original idealism shine through. We should leave them better able to cope with the next seventy years than they were first time around. We are very fortunate in that we now have techniques that enable us to match the original appearance yet do not require excessive maintenance.

How is the Modern Movement different?

It could be argued that the philosophy of conservation of Modern Movement buildings need be no different from that applied to old buildings of any sort. At the other extreme it could be argued that the notion of conservation is against the ideals of the original architects, who would have believed that an obsolete component should be replaced with something better and that an obsolete building should be pulled down.

An instance of the latter situation was a commission to restore a building by Walter Segal. The original client had asked for a playroom for his children, with a design life of ten years. It was then thirty years old and all the significant timbers were suffering from wet rot. The only alternatives were demolition or rebuilding in facsimile. The choice was taken, not without regret, to demolish. A contrasting approach was adopted for Mies van der Rohe's Barcelona Pavilion (built 1928–9, demolished 1930); here the structure was completely rebuilt in 1986, and this is satisfactory if one takes the view that it is the essence, the design ideas, rather than the substance that should be passed on to future generations.

The conservation movement generally, and the Society for the Protection of Ancient Buildings (SPAB) in particular, have placed great value on 'authenticity'. When we look at a stone column, we want to see the hand of the medieval craftsman or the Greek slave, not that element reproduced by a modern builder. But, it can be argued with the buildings of the Modern Movement, it is the design of the building and the spaces within it that are important, and these may be best served by replacing worn elements with new. A much-repaired sixteenth-century timber-framed building may have charm; a much-repaired steel window does not. It therefore makes sense to replace the windows with new. That is certainly what the original architects would have wished.

Conservation theory has also placed value on 'reversibility': the ability to undo repairs that later generations regard as substandard or incorrect. With the buildings of the Modern Movement this approach is no longer reasonable. There is no 'reversible' way of repairing a concrete wall or a steel window.

Architects involved with the conservation of Modern Movement buildings of the 1930s consider themselves lucky that the concrete was painted, so repairs can be hidden. How much less fortunate the conservation architect repairing the *beton brut* of the 1960s!

One thing has become clear about the conservation of Modern Movement buildings: there can be no orthodoxy. There can be no approved way of conserving these buildings. Each building is different and needs thinking about from the starting point of that building, not from the starting point of a particular conservation doctrine. It is this very difference that makes the conservation of Modern Movement buildings so fascinating and so difficult.

Why do people consult architects?

Owners of historic properties that serve their purpose well, that do not have structural problems or unacceptable energy consumption, do not need to consult an architect. It is when there is a problem that the architect is called in. The problem is likely to be more than one of straightforward conservation. If ever the architect's design skills are needed, it is in this situation – one of making a valued historic building also a useful, economically viable, beautiful building that does not squander the earth's resources.

What starts out as a conservation problem soon becomes a design problem. In fact straightforward conservation is usually the easy bit. More difficult issues are poor fit, need for more space, bad detailing, kitchens and bathrooms, and poor energy performance. Let us examine each of these problems in turn.

Poor fit

Most Modern Movement dwellings, such as the apartment buildings Highpoint One and Two, Highgate, London (1933–5 and 1937–8) by Berthold Lubetkin of Tecton, were designed for upper-middle-class families with at least one live-in maid. They were designed with maids' bedrooms at ground level and back lifts so that servants could get to the flat where they worked without being seen by their employers. Even quite large houses would only have one bathroom. Children's quarters could consist of a day nursery with a night nursery off it. Fortunately the service parts of the houses and apartments can be changed without hurting the main living spaces.

The owner will see things from a different angle from the conservation officer acting for the council or for English Heritage. The wise conservation officer will only demand control over those aspects of the design which are really significant to the historic or architectural importance of the building or to maintaining its beauty, and will try to empathise with the owner. The architect has the difficult job of leading the owner into making decisions which respect the building while at the same time nurturing their wishes; fortunately, most Modern Movement buildings are bought by people who love the architecture.

While lifestyle does not change so completely that twentieth-century houses cannot be made to fit in with twenty-first century domestic requirements, when we move outside this sphere changes become more significant. The German Hospital, Hackney, London, found that by 1977 its 1935–6 extension by Burnet, Tait and Lorne was no longer suitable for medical use; listing prevented demolition and it has been converted into flats, keeping the exterior virtually unchanged. The Van Nelle Factory (1926–30) in Rotterdam by Johannes Brinkman with L.C. van der Vlugt and Mart Stam, one of the great pioneer buildings of the Modern Movement, ceased industrial use in the late 1990s. With extensive creative alteration, the coffee and tobacco factory became the 'Van Nelle Design Factory', housing many organisations with a cultural bent; the basis of the building has survived and it looks better than ever.

It must be accepted that there are some buildings for which it is virtually impossible to find new uses. Berthold Lubetkin's Penguin Pool (1933–4) at London Zoo, one of the most popular Modern Movement buildings, is an example. Concepts of the way that we house and display animals have fundamentally changed and the zoo announced in 2004 that the penguins would be housed elsewhere as they mate more successfully in less exposed conditions. Lubetkin would have said 'demolish it and construct something better', but it is difficult to accept this except as a last resort.

The 'lack of fit' may be caused by proposed new uses. At High Cross House, Devon (1930–32), originally designed by William Lescaze for the headmaster of Dartington Hall School, the ground floor was to be converted into an art gallery. The house has large south-facing windows to embrace sun and view, but the pictures to be displayed were extremely light-sensitive, many being watercolours on paper. Fortunately there was space to spare, and it proved possible to turn the north-facing service rooms and the garage into galleries and refurnish the south-facing rooms with their original furniture so that they became as much an exhibit as the pictures (the house reopened as a gallery and archive centre in 1995).

The need for more space

Buildings change over time. It is precisely those changes over time that make historic buildings so interesting; many prefer Durham Cathedral to Salisbury, in spite of the stylistic perfection of the latter. Changes to historic buildings are difficult enough, but additions really do test architectural skills. Adding an extension to a beautiful house is always difficult. In the rare cases where an existing building is such a complete design in itself that nothing can be added without spoiling it, the architect has to come clean and tell the owner that they must either live with the space they have or move.

Unfortunately, postmodernism has undermined our confidence, and teaches us that there are lots of styles and that we have to choose. Extensions to Georgian houses or Tudor cottages should be clearly different from the original. But is the 1930s so far into the remote past that a modern extension must necessarily be different? A white cube with metal windows is still a reasonable way of building today, and a small extension to a concrete house by Tecton makes no attempt to confront Tecton with a newer style, having a finish similar to the main house, but windows of different proportions to show that it is not part of the original design.

In the case of houses by F.R.S. Yorke and Ernö Goldfinger, there was no way of extending the house without doing damage to it. So the solution was to build a separate building in the grounds; in the case of the Yorke house the new building has its floor level set 900 mm below ground level so that it appears very subservient to the main house.

The Cohen house in Old Church Street, Chelsea, London (1935–6), by Erich Mendelsohn with Serge Chermayeff, appeared as a complete composition. But, by building in metal and glass, Norman Foster in the early 1990s added an extension which complements the original by being clearly different in materials and construction, yet it nestles comfortably against the old building so that the whole is greater than the two parts.

With a small extension, as with other interventions to historic buildings, modesty is a great virtue. Theories do not help, as with most architectural problems you have to draw or model it and see if it looks right. Go on drawing and model-making until you are satisfied. Architecture is a visual art. Trust your eyes!

Bad detailing

The architects of the Modern Movement wanted to reject tradition. So out went drips, overhangs, projecting sills and many of the little things that kept buildings dry and in good shape. They learnt after a few years but, perversely, it is the buildings of those first few years that are most loveable.

Above all, these architects had a faith in concrete almost as great as current architects' faith in silicone. Ernö Goldfinger, one of the best of the modern architects, returned from a visit to the Baths of Diocletian in Rome observing that the original concrete was in perfect condition after 1800 years, but that the stonework Michelangelo added 400 years ago was not weathering so well. Alas! The reinforcement that the twentieth century added to concrete caused problems that did not worry Diocletian's builders. The technical problems of dealing with chloride attack, with carbonation and other failings of reinforced concrete, are dealt with elsewhere,[3] but the design issues can be dealt with here.

The architects who built some of the best houses of the Modern Movement not only loved concrete and steel windows; they loved thin concrete and thin window frames. This leaves us with serious difficulties. In the 1930s half an inch (12mm) of concrete cover to reinforcing steel was the recommendation of the code of practice, and three-eighths of an inch (9mm) of cork or Celotex was considered adequate thermal insulation. The regulations now require 50mm cover to steel reinforcement and very substantial thermal insulation. But the attraction of those thin walls remains, and the architectural quality of the building will be lost if the walls become lumpen. So the skilled designer must find ways of keeping the thinness where it shows and ensuring that it is, and will remain, sound.

Most of the houses of this period were designed by architects who loved cubism. Whether this came into architecture from Cezanne's paintings or as a result of Le Corbusier's visit to the Greek islands is immaterial: the love of the pure white cube was strong. In some strange way it was seen as being the appropriate architecture for a machine age. A painter can paint a beautiful pure white cube, but if you build it in the outdoors it will stain and streak. For centuries tops of walls had been protected by overhangs, but this was thrown away as being traditional and impure. The result was buildings that stayed looking good just long enough for the photographs to be taken before they started to deteriorate. We are left with the problem of what to do. Put an overhang on the top, be it ever so small, and the architecture is terribly compromised, to the extent that the architect can but obtain the owner's promise to repaint every year.

One has to be careful how one diagnoses bad details or inadequate structure. Is the problem so serious that it really does matter? The owners of a large listed block of flats in south London, built in 1936, had approached a structural engineer to advise on repairs. The structural engineer investigated the reinforcement in the concrete of the cantilevered balconies and found that it did not comply with current codes of practice. The result of his report was that flats could not be sold, and residents were very worried. Fortunately, another engineer was brought in who ascertained that the reinforcement was in accordance with 1936 codes and was perfectly safe.

Kitchens and bathrooms

Kitchens and bathrooms are often a problem for listed houses built in the 1930s. Everyone will accept that a Georgian mansion or a Tudor cottage may have a modern kitchen and bathroom. But the perception is that a

1930s bathroom can be fine and a 1930s kitchen can be – just about – workable. A more realistic view is that the service rooms of houses are like machines that are changed when it is economic to do so or when a better product comes on the market. But many conservation officers do not see it that way and can drive a would-be owner away by requiring retention of old sanitary fittings, enamel work surfaces or kitchen cabinets. Owners have been known to damage or hide old fittings immediately before a visit from the conservation officer.

In the Lakeshore Drive Apartments, Chicago (1948–51), Mies van der Rohe deliberately left kitchens and bathrooms to the interior designers, taking little interest in their designs. He argued that these were functional spaces, with design partly driven by fashion; they had limited life and would change many times in the life of the building, and so were no more the concern of the architect than was the occupants' furniture.

However, as so often in the conservation of Modern Movement buildings, there are situations where it becomes appropriate to take the opposite view. Only a vandal would lightly cast into a skip a splendid 1930s bathroom's solid sanitary fittings and Vitrolite, but where these are lost it is possible to find appropriate sanitary fittings, and glass similar to Vitrolite is imported from the Czech Republic.

Poor energy performance

When an existing property is being upgraded the Regulations require it to meet present standards of environmental performance, but an exception is made in the case of listed buildings. This omission is a disastrous 'own goal' by the conservation lobby. Public interest in the conservation of old buildings may have passed its peak, but concern with 'sustainability' issues is growing by the hour. For those of us who love old buildings it is very sad that they can be labelled 'licensed polluters'. Responsible practitioners will try to upgrade the energy performance of buildings entrusted to their care, where possible up to the standards of a new building; after all, when installing central heating into an old building, one adopts the same room temperatures one would for a new building. Conservation of old buildings and conservation of the planet are two totally different things. A responsible owner will consider both, but with listed buildings may find that the conservation officer is only concerned with the conservation of buildings.

The houses of the Modern Movement were designed at the time when central heating was becoming common and hence there was a feeling that anything could be kept warm. So it could, at a price! The best of these houses were built with concrete walls and steel windows. Each of these elements lets the heat through in a big way.

The concrete walls may be only 100 mm thick, with, at most, a 20 mm lining of Celotex or cork as a gesture towards thermal insulation. One has the choice of lining the wall with insulation either internally or externally; the trouble is that the original architects liked to show off the thinness of their walls by taking windows to the corners so that the 100 mm wall thickness shows on elevation. The skill is to add the thermal insulation without losing the sharpness of the design. I can best indicate the advantages and disadvantages of adding insulation internally or externally by describing two projects.

The first concerns Torilla, Hatfield, Hertfordshire (1934–5), the first house designed and built by F.R.S. Yorke, situated some twenty miles north of London. The house was empty, and by 1993 the previous owner had allowed the interior to be vandalised as he expected the house to be demolished. The exterior walls are 100 mm thick in situ concrete, with the imperfections of the formwork visible on the exterior. The interior was exposed to the weather and the thin cork lining to the inside face of the external walls had either disappeared or was in an advanced state of decay. Because imperfections to the exterior concrete were felt to be both interesting and attractive, external insulation would not have been appropriate. The exterior was given an extensive anti-carbonation treatment and then painted. Any interesting details to the inside had gone, so there was no problem about lining the inside face of external walls and the underside of the roof with insulation – in this case British Gypsum's thermal board (with integral vapour check) and a plaster skim. This decision threw up an advantage that was not initially foreseen: services could be buried within this 65 mm thick layer. Electric wiring was simply secured to the wall before the lining was applied. Insulated water pipes were run along the inside face of the outside walls and the insulation chased to allow for them.[4]

The second example is a house 'Sunspan' (Figures 8.1 and 8.2) in the outer London suburbs designed by Wells Coates. Previous owners' attempts to keep the water out included tile-hanging the exterior, which was

Figure 8.1 'Sunspan' by Wells Coates, drawings for which were exhibited at the *Daily Mail* Ideal Home Exhibition at Olympia in 1934. The house before restoration.

Figure 8.2 'Sunspan' after restoration.

effective in keeping the water out but disastrous architecturally. Removing the tiles left the outside render in a poor state. In this case the work had to be done while the owner was in occupation, and the budget did not stretch to internal decorations. The entire house was covered with sheets of polystyrene and a render system applied. This had the advantage that the intermediate floor was not a cold bridge (which it becomes when the insulation is applied internally), and the insulation could go over the parapet to join the insulation laid on top of the asphalt roof, giving, together with the roof insulation, a complete insulated jacket to the house without any work having to be done internally. A problem with external insulation occurs at windows, where every jamb becomes a design problem and, when that problem is solved, the windows are further back from the wall face than the original architect envisaged, a situation that can only be righted by bringing the windows forward.

The Modern Movement's love of thin-section Crittall windows is another cause of heat loss. Many of the windows were of ungalvanised steel and these will be at the end of their life and in need of replacement. Replacement windows can be galvanised, powder-coated and weather-stripped for longer life and better thermal performance, but if double glazed the appearance of the section is likely to be a tiny bit heavier. This increase in size of section is particularly noticeable in Art Deco buildings with their love of thin horizontal glazing bars. The alternative to double glazing is to add secondary glazing; this preserves the thinness of the section, but it is awkward to open a window and the reflections, when seen from the outside, look all wrong. Sometimes conservation officers will not allow double glazing, but fortunately some secondary glazing systems do not require fitting into the structure and so listed building consent is not required.

Usually, by changing to a metal glazing bead, it is possible to put a sealed double-glazed unit into a 1930s steel window. But there may be a problem with weight, particularly in the case of French doors, where the load of the additional glass can cause sag and the door then binds on the frame.

Allied to the need to upgrade the thermal performance of a building is the need to improve the acoustic performance. The most common type of Modern Movement building is the private detached house, where at least the occupants only have to listen to their own noise! When there is a need to improve acoustic performance, for example in flats and terraced houses or near motorways and airports, the way of tackling the problem is the same as for any historic building. Sound is reduced by adding mass, by isolation, by separation and by introducing resilience.

Colour

The buildings of the Modern Movement gained their publicity through the splendid photographs that Dell and Wainwright took for the *Architectural Review*. These photographs were in black and white, and the result was that the buildings were assumed to be painted white. It became known as 'The White Architecture', and so persuasive was this belief that even the practitioners came to believe it – on telling an ageing Maxwell Fry that there was blue paint on one of his houses, he replied that someone must have repainted it. He was wrong.

These houses were extremely colourful. The houses of Connell, Ward and Lucas, for example, were green, purple, beige and pink. Amyas Connell said in a speech at the Royal Institute of British Architects, 'Any fool can throw a bucket of white paint over a building.'[5]

When High Cross House at Dartington Hall was restored, the house was all painted white. But in the archives there is a letter from architect to owner stating that the block on the roadside was to be 'blue to match the Devon sky' (an interesting example of designing in keeping with context!). The architect, William Lescaze, wanted not paint but blue render, but the grey of the cement dominated and the result was unsatisfactory, so it was painted white. In the restoration, with so much documentary evidence for blue, it seemed too good an opportunity to miss, so the jolly blue was of our own choosing.

In contrast to Dartington, where colour was added, at F.R.S. Yorke's Torilla it was taken away. The house was originally pink with windows painted duck-egg blue. The owner was a distinguished abstract painter who only works in grey, and totally refused to have any colour on the house, inside or out. The work was done with a grant from English Heritage, who asked for a paint scrape on the windows to ascertain the proper colour. The owner told the operative to keep scraping until he found grey, and fortunately he found it.

While restoring and converting old buildings is a very serious undertaking, paint colour is temporary and gives scope for fun, and for the owner's taste. Our taste in colour is not the same as it was in the thirties and there is no reason to follow precedent unless we want to. The exception to this is the colour of windows. Replacement steel windows are usually powder-coated, a fairly permanent finish, and so the colour should be considered as seriously as other permanent parts of the building.

Conclusions

There are accepted ways of conserving a Queen Anne house or a medieval barn, and conservation officers will usually enforce the norm. For Modern Movement buildings there is, as yet, no norm. DoCoMoMo (DOcumentation and COnservation of buildings, sites and neighbourhoods of the MOdern MOvement) organises international conferences, which have aired many issues, but reached no firm conclusions. This is fortunate for the practitioner, who is thus free to use his or her sensibilities. This is a field where an agreed standard of conservation would be death to the buildings. They were designed with experimentation and innovation in mind and should be looked after with similar values.

The conservation of the best buildings of the Modern Movement calls for skilled architects who are not afraid to use their design skills and for conservation officers who are open minded. It does not call for theories or codes of practice.

Endnotes

1   Francis Reginald Stevens Yorke, *The Modern House in England* (Architectural Press, London, 1937). See also Alan Powers, *Modern: The Modern Movement in Britain* (Merrell, London, 2005).

2   Houses open to the public are 2 Willow Road, London, NW3, by Ernö Goldfinger, owned by the National Trust; High Cross House, Dartington, Totnes, Devon, by William Lescaze, owned by the trustees of Dartington Hall; and The Homewood, Portsmouth Road, Esher, Surrey, by Patrick Gwynne and Wells Coates, owned by the National Trust.

3   Michael Forsyth (ed), *Structures & Construction in Historic Building Conservation* (Blackwell, Oxford, 2007).

4   Chartered Institution of Building Services Engineers, *Guide to Building Services for Historic Buildings: Sustainable services for traditional buildings* (CIBSE, London, 2002), p. 41. Torilla is the only modern building in the guide.

5   *Architect's Journal*, 10 March 1976, p. 467.

9 Conservation and historic designed landscapes

Jonathan Lovie

Background to the conservation of historic parks and gardens

The value of Britain's great designed landscapes, particularly its eighteenth-century parks and gardens, has long been recognised by visitors as one of our greatest contributions to European art and culture. Domestic travellers from Celia Fiennes (1662–1741), to Dr Pococke, Bishop of Osary (1704–65) in the mid-eighteenth century and John Claudius Loudon (1783–1843) in the nineteenth century have left us a steady stream of descriptions of gardens of all kinds which they visited, testifying to the enduring interest of the British in gardening and horticulture. Despite this, official recognition of the importance of designed landscapes as part of our national heritage and environment has only come relatively recently. This is particularly striking as an examination of the early campaigns of what came to be known as the amenity movement reveals a keen awareness of the importance of landscape – both natural and, in the broadest sense, designed – as well as of vernacular buildings; the Commons and Footpaths Preservation Society was formed as early as 1865, while William Morris founded the Society for the Preservation of Ancient Buildings (SPAB) in 1877. The National Trust (now the nation's largest voluntary society with over 3 million members) was founded in 1893, partly to kerb development in the Lake District, and partly to preserve threatened vernacular buildings such as the Clergy House at Alfriston, Sussex, the first historic building acquired by the Trust in 1896.[1] It was not until 1937 that the Trust inaugurated its country houses scheme, developing the sphere of activity for which it has become probably best known.

During the immediate post-war period, the tentative steps taken during the 1930s towards planning for rural areas were renewed and strengthened, with official protection for buildings of historic and architectural importance first being accorded under the Town and Country Planning Act 1944. At the same time, the National Parks and Access to the Countryside Act 1949, and the establishment of the first green belt in 1955, recognised the importance of countryside and open space to the quality of life for millions of urban and suburban Britons. Despite these positive moves, designed landscapes slipped through the net of official recognition;

buildings were now listed, and archaeological remains could be scheduled if they were considered to be of national significance, but there was no provision for recognising the significance of designed landscapes in their own right. Perhaps this was in part because many historic landscapes include elements which were being afforded specific protection; and perhaps other problems seemed more pressing, such as the plight of the country house, highlighted in the 1974 Victoria and Albert Museum exhibition, The Destruction of the Country House, or the spectacular Mentmore sale in 1977.

The work of individuals such as Dame Jennifer Jenkins and Mavis Batey, along with many members of the Garden History Society and other amenity groups which finally led to official recognition of the significance of designed landscapes in our cultural heritage, has been described in detail elsewhere.[2] Thus, the National Heritage Act 1983 empowered the nascent English Heritage to prepare a register of gardens, parks and other land of historic interest. This register differed from the earlier listing of buildings and scheduling of archaeological remains in that no additional statutory powers of protection were attached to the designation, which was, in effect, a means of identifying to planners, owners and other interested parties that a particular designed landscape was considered to be of national significance. The register can thus be seen as part of the essential first step in the process of conserving historic designed landscapes – identification. Identification leads on to the assessment of the significance of a site in its national and local context and, where appropriate, to its designation, which in turn informs the process of preserving, conserving and managing change within the site.

Identifying and understanding historic designed landscapes

Often used as a shorthand for 'historic designed landscapes', the term 'historic parks and gardens' can be misleading. Designed landscapes embrace a far wider range of sites than simply landscape parks and more confined gardens designed for private use. As research and understanding of garden history developed, so public parks and cemeteries came to be seen as significant examples of design, as did specialised institutional landscapes associated with hospitals, asylums or educational establishments. Factories, larger-scale industrial schemes, motorways, airports, Cold War military installations, crematoria and even golf courses – often thought of as the enemy of the landscape park – have their own landscape interest and significance which needs to be properly understood if management is to be effective in conserving interest and distinctive character.

Before any change – and management itself can involve significant change – is contemplated in a designed landscape, it is essential to analyse how the landscape has evolved to be as it is today. Some designed landscapes, such as certain cemeteries, public parks or institutional landscapes, may have a single phase of development, but the vast majority of sites have evolved over a lengthy period and will include elements from several different periods. Even where one phase is more dominant than others, there are often underlying elements incorporated within it. This can be seen in the campus landscape of the University of Birmingham at The Vale, Edgbaston, which was laid out to accompany Casson and Condor's new buildings in the early 1960s, but where features forming part of the setting of the nineteenth-century villas that previously occupied the site were retained and adapted for a new use.

A more typical and more complex example of this layering within the historic development of a designed landscape can be seen at a site such as Combe Abbey in Warwickshire. Here, a Cistercian monastery passed into domestic use in 1539 and gardens, probably of some significance, existed by the early seventeenth century when Princess Elizabeth, daughter of James I, lived in the house for five years. Elaborate formal gardens are shown to the east of the Abbey in an early eighteenth-century bird's-eye view by Knyff and Kip (Figure 9.1),[3] but these were largely removed in the late eighteenth century by Lancelot Brown who laid out a deer park to the west of the house and pleasure grounds and a walled garden to the east. The pleasure grounds were remodelled in the 1860s by the head gardener, William Miller, and in 1897 an elaborate parterre was laid out on terraces to the west of the Abbey (Figure 9.2). When the estate was sold and broken up in 1923, the eastern pleasure grounds ceased to be cultivated, and only the eighteenth-century deer park and the late Victorian terraces west of the Abbey remained in use. From this account, it can be seen that anyone

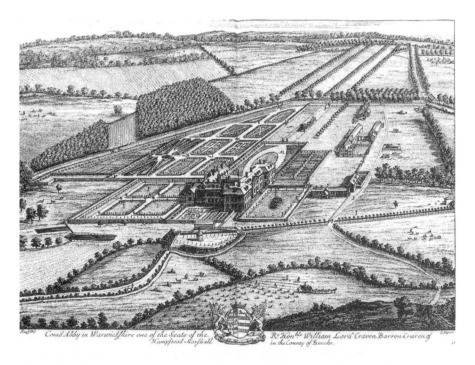

Figure 9.1 Bird's-eye view of Combe Abbey in 1707 by L. Knyff and J. Kip.

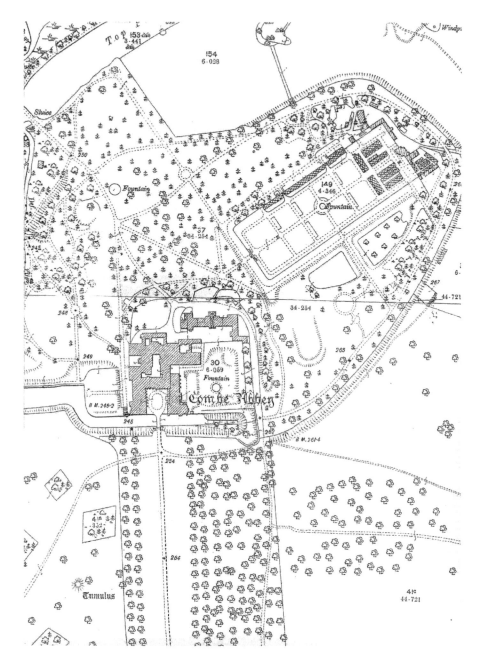

Figure 9.2 Combe Abbey on the Ordnance Survey map of 1903.

trying to frame management proposals for this site without an understanding of its complex historic evolution might conclude that the design emphasis of the landscape is to the west, and that the area to the east might be appropriate for development proposals; whereas, of course, the uncultivated area to the east of the house may retain significant seventeenth-, eighteenth- and nineteenth-century garden archaeology and surviving planting.

The task of researching the development of a historic designed landscape may appear daunting, but a checklist of standard sources will usually help to get the researcher started with any site. The first point of reference for any designed landscape should be map sources, and particularly the late nineteenth- and early twentieth-century editions of the Ordnance Survey. These are usually available in local libraries and in county record offices. Depending upon the age of the site under consideration, other maps, such as the early nineteenth-century Tithe and Enclosure maps, may provide useful information. If the site has been owned by a family that retained an estate archive, this may include surveys and plans, and plans attached to property deeds can also reveal useful evidence for the development of the landscape.

Some sites will have been described at various times by visitors, and the nineteenth-century horticultural press in particular abounded in accounts of gardens up and down the country. A useful point of reference for such material is Ray Desmond's *A Bibliography of British and Irish Gardens* (St Paul's Bibliographies, Winchester, 1984), while his *Dictionary of British and Irish Botanists and Horticulturists* (Taylor & Francis, London, 1994) provides valuable information about many gardeners, designers, nurserymen and owners. County histories such as the authoritative Victoria County Histories, or, for some counties, the volumes prepared by the Royal Commission for Historic Monuments, may prove useful, and many counties now have published studies of their gardens and designed landscapes, while many more studies are in the course of preparation.[4] The county volumes of the Buildings of England series, originally largely written by Sir Nikolaus Pevsner, are steadily being issued in new editions, many of which devote much greater attention to gardens, parks, cemeteries and designed urban spaces than was the case in the first editions. Articles on country houses and gardens published by *Country Life* since 1897 often provide a valuable photographic record as well as useful descriptive text,[5] while many libraries, museums and county record offices hold collections of postcards, photographs and other historic views that should be investigated. In addition, English Heritage's National Monuments Record at Swindon holds extensive collections of photographic material, including the Nigel Temple Postcard Collection and an extensive series of aerial photographs which can be particularly valuable sources of information for designed landscapes.[6] Where a site has a datable phase of development, or is the work of a major designer, one of the specialised studies may give information on the site itself, or indicate useful comparisons.[7]

County record offices contain many estate archives and collections of family papers which may, if the researcher is fortunate, contain important information relating to designed landscapes. In addition to the estate surveys and plans already mentioned, account books, diaries and correspondence can all be good sources of evidence for changes in the gardens and landscape. If the research relates to a public or institutional landscape, the archive material is likely to be found in minute books and ledgers, such as the burial board minutes for a cemetery, or the borough surveyor's records for a public park.

Most English counties now have a county gardens trust, a voluntary association of people with an interest in designed landscapes, which, together with the Welsh Historic Gardens Trust, belong to a national umbrella organisation, the Association of Gardens Trusts.[8] Many of the county trusts have researched historic designed landscapes within their areas, and may be able to provide valuable information. In some cases this material has been published, or deposited at the county record office. The Association of Gardens Trusts is also a member of the Parks and Gardens Data Partnership, a Heritage Lottery-funded project to compile a database of information relating to historic designed landscapes; in the future this may become a useful reference point for researchers.

Once the historic development of a site has been determined, it is important that it is recorded in a clear, well-illustrated report which can be used as the basis for the assessment of the significance of the site, and for framing proposals for ongoing management and conservation. These proposals should be presented in the form of a conservation management plan which, in addition to summarising the development and historic significance of the site, identifies the priorities and constraints arising from both the conservation objectives for the site, and issues such as the economics of farming or forestry which will contribute to the costs of management. Conservation management plans need to be flexible and regularly updated, and readily available – both to owners and managers, and to outside bodies such as planning authorities, statutory consultees or grant-giving bodies – in order that the context in which management decisions and conservation objectives have been reached can be appreciated and understood. Conservation management plans do not need to be lengthy or complex documents; indeed, clarity is essential if they are to be authoritative and well used.

Guidance on the appropriate format for historic landscape conservation management plans has been produced by English Heritage.[9] The Garden History Society has published Advice Notes on *Historic Landscape Assessments* and *Historic Landscape Conservation Management Plans*.[10] These documents form part of a series of fourteen Planning Conservation Advice Notes (PCANs), the first twelve of which consider the issues arising from some of the most commonly proposed types of change in historic designed landscapes ranging from hotel and leisure use, golf course development and telecommunications masts to CCTV, lighting and vehicle access and parking.[11] Other bodies have also published guidance addressing the management and conservation of specific types of sites, such as the joint publication by English Heritage and English Nature on the conservation of cemeteries, *Paradise Preserved* (London, 2002).

Assessment of significance and designation: the national picture

The first edition of the English Heritage *Register of Parks and Gardens of Special Historic Interest* was completed under the direction of Dr Christopher Thacker in 1988. It was always recognised that this would not be a definitive list of nationally significant designed landscapes, and as the result of ongoing research, particularly in the wake of storm damage in 1987 and 1990, sites continued to be identified that merited inclusion on the register. Under Dr Thacker's successors, David Jacques, Harriet Jordan and Sarah Rutherford, a systematic review of potentially registerable sites in each county was undertaken, and the existing register entries were thoroughly upgraded. Thematic studies examined public parks and cemeteries, and detailed research was commissioned on subjects as diverse as detached town gardens, urban squares, medieval deer parks and post-war designed landscapes.[12] All this led to a much better understanding of the surviving heritage of designed landscapes in England and their great diversity; and while the number of nationally significant sites included on the register in 2005 exceeded 1500, resources have not to date enabled the proper assessment of the many other sites which have been identified as being of potentially national significance through the thematic studies and county reviews.

In order to determine whether a site merits national recognition by inclusion on the register, English Heritage assesses it in accordance with a series of published criteria.[13] The first five of these criteria comprise a set of date bands which broadly mirror the main trends in the history and development of gardening and landscape design. Parks and gardens where the design and surviving layout are exceptionally old (early eighteenth century or earlier) are very rare, and having a set of features which is exceptionally old is in itself likely to make a site sufficiently special to be included on the register. Broadly, the more recent the structure of a designed landscape, the more likely it is to have survived, and the more common that type of site is likely to be; so the selection process for such sites must be more rigorous. For a more recent site, such as a Victorian park or a cemetery, to be given national recognition by inclusion on the register, therefore, it must have something particular that makes it 'special'; it might be a site which was influential on contemporary taste through published descriptions and accounts, or it might be the work of a nationally famous designer. It might be a good representative example of a particular type of site, while in another case it might be a site having an association with a nationally significant person or event, or it might form part of a group of sites which, taken together, are of national significance.

As with listed buildings, designed landscapes included on the register are graded to indicate degrees of significance, although it must be remembered that all sites on the register are of national significance. Grade I sites, such as Stowe or Blenheim, are considered to be of international significance and represent some 10% of designated landscapes. Grade II* landscapes are defined as being of 'exceptional historic interest' and make up some 30%, while Grade II sites, some 60% of those on the register, are considered to be of special historic interest. The grading of registered parks and gardens will be changed to conform with the unified designation system introduced by English Heritage, with sites graded at II* being re-graded as Grade I or Grade II. It is unclear what timescale will apply to this process, although from 1 April 2005 any site added to the register was graded in this way.

Each entry on the register consists of a written description of the site as it existed at the time of its assessment for inclusion, together with a description of its historic development and a map delineating the boundary of the area designated as being of special historic interest. The first paragraph of each register description comprises a summary statement of significance, indicating the reason why the site has been assessed as worthy of national recognition. Register entries and boundary maps can be obtained from the National Monuments Record,[14] while those sites that include features such as wood pasture, which are of nature conservation value, are identified on English Nature's parkland information system.[15]

The decision to register a designed landscape has rested historically with English Heritage's Inspector of Parks and Gardens, but under the new designation system, which aims to provide a more holistic approach to the historic environment, four territory-based Heritage Protection Teams will take the lead in making additions to the register. In 2005, English Heritage's garden and landscape casework, including statutory consultations on planning schemes affecting Grade I and II* registered landscapes, was dealt with by regional landscape architects based in the organisation's nine regional offices. Not every region has its own landscape architect; at present they are based in the south-west, south-east, West Midlands, Yorkshire and London, while in the other regions casework is covered by other specialists.

The Register of Parks and Gardens and the planning system

The register has been a valuable means of raising awareness of the significance of historic designed landscapes, and while it imposes no additional statutory controls, through government planning guidance and statutory duties of consultation it has helped to establish parks and gardens as 'heritage assets' of national significance, and to bring them into the planning system.

Planning Policy Guidance Note 15 (PPG15), to date the only statement of government policy on planning and the historic environment, provides clearly worded advice to planning authorities, owners and developers:

> Local planning authorities should protect registered parks and gardens in preparing development plans and in determining planning applications. The effect of proposed development on a registered park or garden or its setting is a material consideration in the determination of a planning application. Planning and highway authorities should also safeguard registered parks or gardens when themselves planning new developments or road schemes.[16]

PPG15 not only establishes registered parks and gardens as a material consideration in the planning process, but also clearly indicates that planners should assess the likely impact of any proposals on the setting of the registered site. This advice has been confirmed by case law and a series of decisions by the Planning Inspectorate; and it should be borne in mind that the setting of a registered landscape is not confined to the land immediately adjoining the boundary of the designated site – what might be termed the 'essential setting'[17] – but may extend for some distance beyond the site. It is clear that proposals for a telecommunications mast, for example, on a high, visually prominent site on the boundary of a registered site would have an adverse impact which the planning authority should consider in its determination of the application; similarly, the impact of a proposed group of wind turbines some distance outside a registered site, but still in a visually prominent and intrusive position, would be a material consideration. The adverse impact of such a scheme would be increased further if the site impinged on a designed vista radiating out from the core of the designed landscape. Such views are often an integral component of the intended landscape experience, and their loss can have a serious impact on the integrity of the site and its special historic interest.

Many historic designed landscapes include other features of historic interest such as archaeological remains or listed structures. PPG16 (*Archaeology and Planning*) provides helpful guidance on the conservation of archaeology through the planning system, while advice on the conservation of areas of historic designed landscapes now in agricultural use is provided by the Department for Environment, Food and Rural Affairs (Defra).[18] It is important to remember that some historic parks and gardens themselves survive only as archaeological features; Lyveden New Bield, Harrington and Holdenby, all in Northamptonshire, are good examples of this type of site, which is clearly vulnerable to damage through inappropriate agricultural practice as well as through more obvious development. Where a historic landscape is itself a scheduled ancient monument, or includes such a feature within its boundaries, consent is required for any development which might affect it. Similarly, where a landscape includes a listed building, such as a country house, other adjacent structures such as terraces, steps, walls or gateways will be considered to form part of the curtilage of the listed building, and will have the benefit of a similar level of protection from inappropriate change or development.

Since its publication in 1994, PPG15 has proved to be an effective statement of planning policy for the historic environment, balancing the need for conservation with allowing appropriate change and development. The government announced its intention to rewrite PPG15, along with other Planning Policy Guidance Notes, as part of a series of Planning Policy Statements (PPS). PPG15 remains one of the few guidance notes not to be revised. The government's timetable for this work remains unclear, but it would be tragic for the historic environment as a whole if the aspiration and breadth of vision of PPG15 were to be lost.

Although PPG15 deals principally with the overarching national planning framework, it makes clear that the historic environment has a very local aspect:

> The physical survivals of our past are to be valued and protected for their own sake, as a central part of our cultural heritage and our sense of national identity . . . Their presence adds to the quality of our lives, by enhancing the familiar and cherished local scene and sustaining the sense of local distinctiveness which is so important an aspect of the character and appearance of our towns, villages and countryside.[19]

Planning authorities are instructed to include policies to protect registered parks and gardens in their structure plans, local plans and unitary development plans (soon to be replaced by regional spatial strategies and local development frameworks); and most do now include such policies. The Garden History Society has published advice on appropriate wording for these policies,[20] and in their roles as statutory consultees, both the Society and English Heritage seek to comment on as many plan policies as possible at their draft stage. Given that Section 54(A) of the Planning and Compensation Act 1991 gives considerable weight to the development plan in the planning process, these policies are almost a de facto form of statutory protection for registered sites, and their importance cannot be overestimated.

Many local plans also now include a policy that recognises the existence and significance of other historic designed landscapes which have not been included on the English Heritage register. In some policies, the protection afforded to these sites is as great as that afforded to registered parks and gardens, and indeed, the Planning Inspectorate has indicated in its reports that it takes the view that there are no reasonable grounds for applying different policies to registered and unregistered sites.[21] Many of these 'local lists' included in local plans will have been drawn up with the advice of local amenity groups with specialist expertise, such as the county gardens trust; in the best examples, the lists are backed up with robust documentation and maps defining the areas of special interest, which in turn can be incorporated into the plan's proposals map. It is clearly desirable that an explanatory memorandum should be added to local lists indicating that more sites may be added to the list during the lifetime of the plan as knowledge and understanding extend. Such a memorandum should also refer to the planning authority's statutory duty under Department of the Environment Circular 9/95 to consult English Heritage on planning schemes affecting Grade I and Grade II* sites, and the Garden History Society on schemes affecting all registered sites regardless of grade.[22] It follows that it is desirable that an intention to consult local amenity bodies such as the county gardens trust on sites of local significance also be indicated in the plan.

Wherever possible, lists of locally significant designed landscapes should be included on the county Sites and Monuments Record or Historic Environment Record. This not only helps to identify the significance of the site should a planning issue arise, but also forms a relationship between the designed landscape and other archaeological and built features that are included on the record.

Lists of locally significant designed landscapes are a valuable opportunity for members of the local community to identify and seek to conserve elements of the 'cherished local scene' which give distinctiveness to where they live. Another vital, locally generated conservation tool is the designation by planning authorities of conservation areas. Section 69 of the Planning (Listed Buildings and Conservation Areas) Act 1990 imposes a duty on local planning authorities to designate as conservation areas any 'areas of special architectural or historic interest the character or appearance of which it is desirable to preserve or enhance'. Many conservation areas include within them a historic designed landscape, which may or may not be on the English Heritage register. In urban areas, the conservation area might comprise a distinctive nineteenth-century residential development around a public park; or it might include a cemetery, such as the Key Hill and Warstone Lane cemeteries in the Birmingham Jewellery Quarter, which have a special historic connection and relevance to the development of the area. In a rural context, a conservation area may comprise a designed landscape and an associated estate village, or in some cases it may comprise only the designed landscape and its associated structures, as does the Rushton Hall conservation area in Northamptonshire. Once again, PPG15 provides clear guidance:

> Designation may well . . . be suitable for historic parks or gardens and other areas of historic landscape containing structures that contribute to their special interest . . .[23]

Many conservation areas include a wide range of disparate designed landscape elements. For example, a conservation area in the centre of Rugby includes within it the Regency landscape designed to accompany the buildings of Rugby School, several nineteenth-century villa gardens (at least one of which is unchanged since 1855), an area of designed open space forming the focal point of a late nineteenth-century villa development, and a late eighteenth-century serpentine garden wall which survives from an earlier manor house. None of these landscape features is likely to merit national recognition, but each is individually significant in a local context, and collectively, particularly in their relationship to the other elements of the historic environment, they make a significant contribution to the distinctive character of the locality.

In its guidance on the designation of conservation areas, PPG15 places a duty on planning authorities both to undertake an initial assessment of the area to be designated, and to make regular reviews of existing conservation areas.[24] These processes are essential, not only as a means of explaining to the local community the reasons for designation and identifying the special interest and character which it is intended to conserve, but also to monitor the effectiveness of the designation and the impact of changes during a given period. Further guidance on the designation and management of conservation areas is provided by English Heritage in its publication *Conservation Area Practice* (1995).

Increasingly, planning authorities are recognising that development pressure in areas such as residential suburbs, which have traditionally had a lower density of development, is leading to a significant change of character. An area characterised by detached Edwardian villas set in spacious gardens rapidly loses this character if the villas begin to be demolished and the gardens developed with smaller houses or apartments, or even when front gardens are given over to car parking. Sometimes it is possible to designate such places as conservation areas, but in other cases the planning authority may find it more appropriate to issue supplementary planning guidance, setting out the limits and form of development that the authority will consider appropriate in a particular area. Supplementary planning guidance is thus a most helpful tool when seeking to preserve designed landscape features such as domestic gardens or boundary treatments which, while clearly not of national significance, nevertheless make a significant contribution to the valued character of a locality.

Enabling development and historic landscapes

During the 1980s and 1990s it became increasingly common to justify development proposals on the grounds that, if implemented, the scheme would benefit a neglected or damaged 'heritage asset' such as a derelict listed building. Many of the schemes were, however, ill thought through and resulted in more damage than benefit. Designed landscapes of national or local significance were particularly vulnerable to damage through development which claimed to fund the repair of listed buildings; a scheme of this kind in the 1980s allowed the renovation of a Grade I listed building at the expense of destroying, through development, the site of Joseph Addison's early eighteenth-century garden at Bilton Hall, Warwickshire – potentially a highly significant site for the understanding of the evolution of landscape design in the eighteenth century. Disasters such as this arose because of the failure of all those involved in the planning process to recognise the complex relationships which exist between the different elements of the historic environment, and that in some circumstances the special historic significance of the designed landscape may equal, or exceed, that of a listed building.

Concern at the damage being caused to the historic environment through enabling development has led to a clear presumption against permitting development that harms the overall historic environment, even if gains for a specific element are being offered. Permission should only now be granted if the proposals will not damage the historic environment, and where the applicant can demonstrate that, on balance, the benefits of the scheme to the historic environment outweigh any disbenefits. A valuable policy statement[25] on enabling development published by English Heritage in 1999 and the subsequent guidance on the assessment of applications for enabling development[26] make it very clear that anyone seeking permission for such a scheme must take a holistic view, and be able to demonstrate a good understanding both of the site that it is sought to repair, and its wider historic setting.

Historic landscapes and the reform of the designation system

A more holistic approach to, and understanding of, the historic environment is seen by the government and English Heritage as one of the key factors in ensuring its better management and conservation. At the same time, as a result of dialogue and consultation, the understanding of what constitutes the historic environment has been broadened out to include not only the obvious sites that are of national significance, but also many of those sites that give different places their distinctive character. As part of this process, the government has embarked on a process of reforming the designation, with the intention that in due course the present systems of listing, scheduling and registering historic assets will be integrated into a unified Register of Historic Sites and Buildings for England. This new designation will include a local section which will incorporate local authority designations such as conservation areas, and, presumably, lists of locally significant historic designed landscapes. A unified heritage consent system will be introduced, and provision for the introduction of statutory management agreements for complex historic sites will be made.[27] This is an interesting development which will undoubtedly affect many nationally designated designed landscapes that include listed buildings and archaeological remains within their boundaries.

The first steps towards introducing the unified designation system were taken on 1 April 2005, and it is too early to know how the new system will work in practice. It is clear, however, that the underlying philosophy and the more holistic approach to the historic environment are valuable in encouraging good management and conservation.

Conclusion

Historic designed landscapes are one aspect of the wider historic environment, and frequently include within their boundaries other features of historic or ecological interest. In trying to manage and conserve historic landscapes, it is essential first to understand the nature of the designed landscape, in terms of both its historic development and the elements which survive on the ground. This process informs an assessment of the relative significance of the constituent areas and elements of the site, which in turn will inform a conservation management plan, the object of which is to determine how best the special qualities and character of the historic environment can be conserved. Only after this process has been undertaken, and a thorough understanding of the site and its significance has been achieved, can it be determined what level, if any, of change and development is acceptable and compatible with the long-term conservation of the site.

Whether we like it or not, the conditions and circumstances of the early twenty-first century will not allow us to preserve the historic environment in aspic. The pressures for development and change are enormous, and as English Heritage itself has stated, the role of all those involved in the conservation of the historic environment is to be 'positive managers of change'. The delicate balancing act which must, at all costs, be performed, is to allow the historic environment to continue to develop where appropriate, but at the same time to conserve those qualities which make it special and for which it is enjoyed and cherished, at both a national and local level, by countless thousands, now and in the future.

Curiously, if we were to seek an expression of the philosophy that should underpin our approach to the conservation of historic designed landscapes, we could do worse than consider words addressed by Alexander Pope to Lord Burlington some 300 years ago:

> All must be adapted to the Genius and the Use of the Place and the Beauties not forced into it, but resulting from it . . .[28]

Sources of advice and support

Managing change in historic designed landscapes can appear a daunting undertaking, but there are many bodies offering advice and support to individuals and professionals working in the field. Some useful contacts would include:

Department for Environment, Food and Rural Affairs (Defra): advice on agri-environment schemes which include conservation of historic parklands and component features: www.defra.gov.uk

English Heritage: specific advice on historic parks and gardens and on conservation management plans: www.english-heritage.org.uk/parksandgardens

Forestry Commission: advice on grants for park woodland and shelterbelts: www.forestry.gov.uk

The Garden History Society: www.gardenhistorysociety.org

County Gardens Trusts: www.gardenstrusts.org.uk

National Amenity Societies: Georgian Group, Victorian Society, Society for the Protection of Ancient Buildings (SPAB), Ancient Monuments Society, Civic Trust, Twentieth-Century Society and Council for British Archaeology – all can provide specialist advice in their own fields and belong to the Joint Committee of the National Amenity Societies: www.jcnas.org.uk

Local authorities: a list of county Sites and Monuments Records and Historic Environment Records is available from the Association of Local Government Archaeological Officers' website: www.algao.org.uk

Endnotes

1   Merlin Waterson, *The National Trust* (BBC, London, 1994), p. 41.
2   Mavis Batey, David Lambert and Kim Wilkie, *Indignation!: The campaign for conservation* (Kit-Cat Books, London, 2000); John Anthony, 'Protection for historic parks and gardens', *Garden History* **24**, 1 (Summer, 1996), 3–7; English Heritage, *Conservation Bulletin* **49** (Summer 2005), 33–4.
3   L. Knyff and J. Kip, *Britannia Illustrata . . .* (D. Mortier, London, 1707), pl. 47.
4   Examples might include Todd Gray, *The Garden History of Devon* (University of Exeter Press, Exeter, 1995); Stewart Harding and David Lambert, *Parks and Gardens of Avon* (Gardens Trust, Bristol, 1994); Paul Stamper, *Historic Parks and Gardens of Shropshire* (Shropshire Books, Shrewsbury, 1996); Timothy Mowl, *Historic Gardens of Dorset* (Tempus, Stroud, 2003).
5   A regularly updated cumulative index of these articles is published by *Country Life*. The magazine's Picture Library holds copies of published and unpublished photographs, while back copies are held by many libraries and by the National Monuments Record (NMR), Swindon.
6   Stroud, repton. See www.english-heritage.org.uk/nmr for information on the National Monuments Record and its collections.
7   Random examples include Roy C. Strong, *The Renaissance Garden in England* (Thames and Hudson, London, 1979); David Jacques, *Georgian Gardens: The reign of nature* (Batsford, London, 1983); Brent Elliott, *Victorian Gardens* (Batsford, London, 1986). Studies of designers such as Dorothy Stroud, *Capability Brown* (Faber & Faber, London, 1975) and *Humphry Repton* (Country Life, London, 1962), or Stephen Daniels, *Humphry Repton: Landscape gardening and the geography of Georgian England* (Yale University Press, New Haven and London, 1999), are also helpful, as are studies of particular types of site, such as Hazel Conway, *People's Parks* (Cambridge University Press, Cambridge, 1991) which examines public parks.
8   See www.gardenstrusts.org.uk
9   English Heritage, *Conservation Management Plans for Restoring Historic Parks and Gardens* (consultation draft, 2001).
10 Garden History Society, PCAN13 and PCAN14 (Garden History Society, London, 2005).
11 Garden History Society, PCAN1: *Change of Use*; PCAN2: *Hotel and Leisure Development*; PCAN3: *Extension of Educational/Institutional Establishments*; PCAN4: *Executive Housing*; PCAN5: *Golf*; PCAN6: *Vehicle Parking and Access*; PCAN7: *Treatment of Boundaries and Entrances*; PCAN8: *Telecommunications Masts*; PCAN9: *Development of Domestic Amenities*; PCAN10: *CCTV and Lighting*; PCAN11: *Development in the Setting of Historic Designed Landscapes*; PCAN12: *Evaluation of New Landscape Features*; PCAN13: *Briefs for Historic Landscape Assessments*; PCAN14: *Management Plans (including statements of significance)*, (Garden History Scociety, London, 2005).
12 Copies of these thematic studies and the county reviews are held by English Heritage, London.
13 The criteria are published on the English Heritage website www.english-heritage.org.uk/parksandgardens
14 www.english-heritage.org.uk
15 www.wapis.org.uk
16 PPG15, para. 2.24.
17 It should be noted that the mapping for the Welsh Register compiled by Cadw defines the 'essential setting' of the designated sites; however, case law has established that 'setting' is not confined to this delineated area, but in certain circumstances extends significantly further.

18 English Heritage, *Farming the Historic Landscape* (English Heritage, London, 2005).

19 PPG15, para. 1.1.

20 Garden History Society, *Advice on the Protection of Historic Parks and Gardens in Development Plans* (Garden History Society, London, 2000).

21 North Somerset Council, *Woodspring Local Plan to 2001: Inspector's report* (North Somerset Council, North Somerset, 1998).

22 For purposes of pre-application consultations and statutory consultations, the Garden History Society can be contacted at conservation@gardenhistorysociety.org.

23 PPG15, para. 4.6.

24 PPG15, paras 4.4, 4.5.

25 English Heritage, Policy Statement, *Enabling Development and the Conservation of Heritage Assets* (English Heritage, London, 1999).

26 English Heritage, *Enabling Development and the Conservation of Heritage Assets* (English Heritage, London, 2001).

27 English Heritage, *Listing is Changing* (English Heritage, London, 2005).

28 Alexander Pope, *Epistles to Several Persons*, IV (London, 1731).

10 International standards and charters

Philip Whitbourn

After a generation of operating relatively sophisticated legislation designed to protect our historic environment in the UK, it may seem somewhat strange to reflect upon the very different situation that obtained in the year 1964. At that time, the concept of conservation areas did not exist, nor did the system of listed building consents, and the financial penalty for demolishing a historic building without notice was derisory. Yet that was the situation in the UK when the second International Congress of Architects and Technicians of Historic Monuments met in Venice in May 1964 and approved a text that we now know as the 'Venice Charter' and as the basis of modern conservation.

The congress in Venice was, however, mindful of the contribution made towards the development of an extensive international movement at a conference held in Athens a generation earlier. At this Athens gathering certain basic principles were defined for the first time. In particular, under the **Athens Charter** of October 1931, although the difficulty of reconciling public law with the rights of individuals was noted, recognition was given to a certain right of the community in regard to private ownership. Also, it was recommended that the surroundings of ancient monuments should be given special consideration, and that buildings should be used for a purpose which respects their historic or artistic character. A current issue in 1931 was the restoration of the Parthenon, and one of the sessions of the Athens conference was held on the Acropolis. At this, unanimous approval was given to the reinstatement of the southern peristyle of the Parthenon by re-erecting original fragments that had fallen from the monument, a process technically known as anastylosis (Figure 10.1).

'People are becoming more and more conscious of the unity of human values and regard ancient monuments as a common heritage,' read the preamble to the **Venice Charter** of 1964. The document went on: 'The common responsibility to safeguard them for future generations is recognised. It is our duty to hand them on in the full richness of their authenticity.' Accordingly, it was agreed that certain guiding principles should be laid down on an international basis, with individual countries applying these principles within the framework of their own culture and traditions. Article 3 of the Venice Charter made it clear that the intention in conserving and restoring monuments was to safeguard them no less as works of art than as historical evidence. Article 11 followed up this point by seeking respect for the valid contributions to a building of various periods. The removal of

Figure 10.1 View of the Parthenon, Athens, before the central part of the southern peristyle was reinstated by a process of anastylosis.

superimposed work of different periods to reveal an underlying state was seen as justified only when that which was to be removed was of little interest and when the material brought to light was of great historical, archaeological or aesthetic value. The importance of precise documentation was emphasised in Article 16, which advocated the production of illustrated analytical and critical reports. Also encouraged were the publication of records and the placing of them in the archives of public institutions, where they could be made available to research workers.

In the following year, 1965, participants from the same twenty or so countries that had drafted the Venice Charter met in Poland to give added effect to the charter by the formation of the **International Council on Monuments and Sites** (ICOMOS). This is now a 5000-strong body of conservation professionals with national limbs in upwards of a hundred countries, and it holds worldwide general assemblies for its members at about three-yearly intervals. In the course of several of these general assemblies further international charters have been adopted on different aspects of conservation, the charters usually taking their names from the venue of the particular international gathering. Also, ICOMOS has some twenty or so specialist committees concerned with particular conservation disciplines, which bring their collective expertise to bear upon the text of the international charters.

The **Florence Charter** on the presentation of gardens was drawn up by the ICOMOS International Committee in Florence in May 1981 and was registered by ICOMOS in the following year as an addendum to the Venice Charter covering the specific field concerned. While advocating the preservation of historic gardens within the spirit of the Venice Charter, the point was made that gardens are living monuments whose constituents are perishable and renewable. Thus it was pointed out that their appearance reflected the perpetual balance between the cycle of the seasons, the growth and decay of nature, and the desire of artists and craftsmen to keep them permanently unchanged.

The **Washington Charter** for the conservation of historic towns and urban areas was adopted by ICOMOS in October 1987 to complement the Venice Charter, in the face of perceived threats, damage, degradation and destruction by the impact of urban development following industrialisation in societies everywhere. To be effective, it was thought necessary for conservation to be an integral part of planning policy, and the encouragement of the involvement of residents was also advocated. 'Conservation plans,' read Article 5 of the Washington Charter, 'must address all relevant factors including archaeology, history, architecture, techniques, sociology and economics.' The charter concluded with a statement in Article 16 that 'specialised training should be provided for all those professions concerned with conservation'.

The **Lausanne International Charter** for Archaeological Heritage Management was adopted in 1990 and opened with the following introductory statement: 'It is widely recognised that a knowledge and understanding of human societies is of fundamental importance to humanity in identifying its cultural and social roots.' Article 1 of the charter went on to define archaeological heritage as 'that part of the material heritage in respect of which archaeological methods provide primary information'. This was intended to include the various vestiges of human existence, places relating to manifestations of human activity, abandoned structures and associated portable cultural material. In Article 2 of the Lausanne Charter it was pointed out that the archaeological heritage is a fragile and non-renewable cultural resource. Policies for its protection should thus constitute an integral component of policies relating to land use, development and planning, as well as of educational policies. An overriding principle that 'the gathering of information about the archaeological heritage should not destroy any more archaeological evidence than is necessary for the protection or scientific objectives of the investigation' was set out in Article 5 of the charter. Non-destructive techniques, such as aerial survey and sampling, were therefore encouraged wherever possible, in preference to total excavation. Article 7 made the point that the presentation to the general public of archaeological heritage was an essential method of promoting an understanding of the origins and development of modern societies, as well as an important means of promoting its protection.

At its General Assembly held in Colombo, Sri Lanka, in the summer of 1993, ICOMOS adopted **International Guidelines on Education and Training** in the Conservation of Monuments, Ensembles and Sites. The document recognised that many different professions needed to collaborate within the common practice of conservation. Thus it was stated that conservation courses needed to be multidisciplinary and to produce conservation professionals with the ability to:

1. read a monument, ensemble or site and identify its emotional, cultural and use significance
2. understand the history and technology of monuments, ensembles or sites in order to define their identity, plan for their conservation, and interpret the results of this research
3. understand the setting of a monument, ensemble or site, its contents and surroundings, in relation to other buildings, gardens or landscapes
4. find and absorb all available sources of information relevant to the monument, ensemble or site being studied
5. understand and analyse the behaviour of monuments, ensembles and sites as complex systems
6. diagnose intrinsic and extrinsic causes of decay as a basis for appropriate action
7. inspect monuments, ensembles and sites, and make reports intelligible to non-specialist readers, illustrated by graphic means such as sketches and photographs
8. know, understand and apply UNESCO conventions and recommendations, and ICOMOS and other recognised charters, regulations and guidelines
9. make balanced judgements based on shared ethical principles, and accept responsibility for the long-term welfare of cultural heritage
10. recognise when advice must be sought and define the areas in need of study by different specialists, e.g. wall paintings, sculpture and objects of artistic and historical value, and/or studies of materials and systems
11. give expert advice on maintenance strategies, management policies and the policy framework for environmental protection and preservation of monuments and their contents, and sites
12. document works executed and make same accessible
13. work in multidisciplinary groups using sound methods
14. be able to work with inhabitants, administrators and planners to resolve conflicts and to develop conservation strategies appropriate to local needs, abilities and resources.

This checklist now underpins the structure of conservation courses generally. In addition, the document suggests that specialist courses may include a library and documentation centre providing reference collections, access to computerised information networks, a range of monuments and sites within a reasonable radius, and teaching facilities with audio-visual equipment.

The ICOMOS General Assembly in Sofia, Bulgaria, in October 1996 adopted a set of **Principles for the Recording of Monuments, Groups of Buildings and Sites**. The document took as its starting point Article 16 of the Venice Charter regarding the production of records and reports, and was directed alike at professionals, managers, politicians, owners, administrators and the general public. Section 1.2 of the document concerned the appropriate level of detail for particular purposes, while section 3.2 touched upon various methods of recording, such as measured plans, photo-grammetry, rectified photography and written descriptions and analyses.

Also ratified at the Sofia General Assembly in 1996 was the International Charter on the Protection and Management of **Underwater Cultural Heritage**. In introducing this subject, the document emphasised that by its very character, underwater cultural heritage was largely an international resource, resulting from trade and travel. Ships and their contents could thus be lost at a distance from their origin or destination. Article 1 of the charter set out certain fundamental principles including consideration of preservation in situ as a first option; the encouragement of non-destructive techniques, the avoidance of disturbance to human remains or venerated sites, and the need for adequate documentation. Article 2 provided a helpful checklist of matters to be taken into account in designing an investigation project. These included techniques to be employed; the anticipated funding; the timetable for completion; arrangements for collaboration with museums; health and safety considerations; and report preparation and documentation.

At the ICOMOS General Assembly held in Mexico in October 1999 two more international charters were adopted, together with a set of Principles for the Preservation of **Historic Timber Structures**. The aim of these principles was to define basic and universally applicable principles and practices for the preservation of historic timber structures, with due respect to their cultural significance. Again, Article 16 of the Venice Charter served as the starting point, with section 1 of the International Timber Charter stressing the importance of carefully recording the condition of a structure before any intervention. Section 2 advocated a thorough and accurate diagnosis of the condition and causes of decay. In section 6 minimum intervention in the fabric was held up as an ideal, with the principle that as much as possible of historic fabric should be retained. This, section 7 emphasised, applied to items such as roofs, floors, doors, windows, infill panels and weatherboarding, as well as to structural members. New parts, the document suggested, should be discreetly marked so that they could be identified later.

One of the international charters adopted by ICOMOS at its 1999 General Assembly was that on the **Built Vernacular Heritage**. Vernacular building was defined as 'the traditional and natural way by which communities house themselves'. That type of heritage was described as utilitarian but at the same time possessing interest and beauty. Thus it was seen as important in its expression of the culture of a community, its relationship with its territory and, at the same time, the expression of the world's cultural diversity. However, because of the homogenisation of cultures and global socio-economic transformation, vernacular structures around the world were considered extremely vulnerable. The continuity of traditional building systems and of craft skills associated with the vernacular was seen as fundamental to conservation, and thus needed to be retained, recorded and passed on to new generations.

Also in Mexico in 1999 the opportunity was taken to adopt an updated **International Tourism Charter**. An international conference held in

Brussels in November 1976 had concluded with a Charter of Cultural Tourism, which had been endorsed by some eighteen international agencies and associations, and this had been adopted by ICOMOS at its General Assembly in Rostok, Germany, in May 1984. However, it was increasingly recognised that tourism and leisure had become major social and economic forces and one of the world's largest sources of employment. Moreover, national governments had committed themselves to the concept of sustainable development, as expressed in the Earth Summit held in Rio de Janeiro in 1992. Thus one of the principles of the 1999 charter was that cultural heritage should be managed in a sustainable way for present and future generations, recognising that the relationship between heritage places and tourism was dynamic and might involve conflicting values.

As the 1984 International Charter of Cultural Tourism became increasingly in need of updating in the 1990s, the UK National Committee of ICOMOS produced its own Statement of Principles for the Balanced Development of **Cultural Tourism** in 1997. This followed major conferences organised by ICOMNOS-UK at Canterbury in 1990 and at Bath in 1995. The latter conference, on the theme 'Historic Cities and Sustainable Tourism' included participants from Finland, France, Hungary, Malta, the Netherlands, Poland, Portugal and Slovakia, as well as from various parts of the UK. The outcome was the commending of **seven principles** for the balanced development of cultural tourism. These were:

1. The environment has an intrinsic value which outweighs its value as a tourism asset. Its enjoyment by future generations and its long-term survival must not be prejudiced by short-term considerations.
2. Tourism should be recognised as a positive activity with the potential to benefit the community and the place as well as the visitor.
3. The relationship between tourism and the environment must be managed so that it is sustainable in the long term. Tourism must not be allowed to damage the resource, prejudice its future enjoyment or bring unacceptable impact.
4. Tourism activities and developments should respect the scale, nature and character of the place in which they are sited (Figure 10.2).
5. In any location, harmony must be sought between the needs of the visitor, the place and the host community.
6. In a dynamic world some change is inevitable, and change can often be beneficial. Adaptation to change, however, should not be at the expense of any of these principles.
7. The tourism industry, local authorities and environmental agencies all have a duty to respect the above principles and to work together to achieve their practical realisation.

Other useful ICOMOS-UK doctrinal texts include Guidelines on **Archaeology in the Management of Gardens, Parks and Estates**, which were adopted in October 1999. In this document it was accepted that the restoration of historic gardens could promote an enhanced appreciation of them historically and visually. However, it was considered that presentational aims should be subordinate to conservation ones. Techniques

Figure 10.2 Royal Crescent, Bath, showing the unacceptable impact of tourist buses before their re-routing to a more respectful distance.

touched upon in checklist form included plant surveys, analysis of garden structures, earthworks recording, historical ecology, aerial and other photography, geophysical surveys such as resistivity, archaeological trial trenching and augering, and soil chemistry.

At the 14th General Assembly in October 2003 at Victoria Falls, Zimbabwe, ICOMOS ratified Principles for the **Analysis, Conservation and Structural Restoration of Architectural Heritage**, where the basic concepts of conservation are presented, and Principles for the **Preservation and Conservation/Restoration of Wall Paintings**.

Particular national committees of ICOMOS have also adopted various doctrinal texts, and almost certainly the best known and the most widely used of these is the Australia ICOMOS Charter for Places of Cultural Significance, known as the **Burra Charter**. This sought to build upon the Venice Charter, and was first adopted by Australia ICOMOS in August 1979 at the historic South Australian mining town of Burra. The text has been constantly updated and revisions were adopted in 1981, 1988 and 1999. Also, a very helpful illustrated version was produced, which fleshed out the otherwise necessarily somewhat drier 'Articles' of a doctrinal text. One of the many virtues of the Burra Charter was the three-stage 'Burra Charter Process' which set out a logical sequence of investigations, decisions and actions. The first stage was to understand the significance of the place. This normally entailed gathering and recording documentary, physical and other information about the place, assessing its significance, and preparing a 'statement of significance'. The second stage involved the development of policy from that statement of significance. For this, information needed to be gathered about the factors affecting the future of the place, such as physical condition and the availability of resources. Policy could then be developed by identifying options and testing these against their impact upon the significance, the concluding element of stage 2 being the preparation of a statement of policy. Stage 3 was to manage the place in accordance with that policy, developing and implementing strategies by means of a management plan where appropriate, monitoring and review being an important element of such a plan. For the purposes of the Burra Charter, 'cultural significance' means aesthetic, historic, scientific, social or spiritual value for past, present or future generations, and it can be embodied in a place itself, its fabric, setting, use, associations, meaning, records, related places or related objects. The charter sought to set a standard of practice for those who provide advice, make decisions about, or undertake works to places of cultural significance, be they historic, indigenous, or natural places with cultural values. The clarity of the logical sequence of the three-stage Burra Charter process – (1) **understand significance**, (2) **develop policy**, (3) **manage** – is one that can be thoroughly commended to all concerned with sound conservation practice.

11 Conservation legislation in the United Kingdom: a brief history

Colin Johns

The historic environment is today an integral part of the planning process with a range of legislation setting out the way in which protection of historic buildings and areas is to be achieved. There is continuing debate concerning the efficiency and effectiveness of the legislation but it is generally believed that with some notable exceptions designation has been effective in protecting those aspects of historic buildings and monuments that are of the greatest significance. It has, however, been less effective in relation to historic places including conservation areas.

In 2000 English Heritage was asked by the government to lead a review of policies relating to the historic environment in England and came to the conclusion that 'the way in which the legislation has developed piecemeal over 120 years is frustrating for regulators and regulated alike'. The report considered that since 1882, when the first legislation was passed, a complicated regulatory system had developed in England. There is a designation system covering scheduled monuments, Grade I, II* and II listed buildings, conservation areas, Grade I, II* and II registered parks and gardens and registered battlefields. Owners of complex sites may find themselves subject to a confusing plethora of different regulatory regimes operated by different authorities.

Although some may consider that the legislation has developed piecemeal, in reality this is simply a reflection of the social, economic and cultural climate of the time. It was therefore inevitable that the legislation should have evolved, and will continue to evolve, to respond to the aspirations of the day. An examination of the legislation from 1882 illustrates that there were a number of milestones along the way which have had a direct effect on the legislators. In some cases the legislation was an integral part of government thinking but it was affected additionally by external sources. It was also influenced by public opinion, particularly in the latter part of the twentieth century. Understanding how and why the legislation was enacted and the key influences is significant in the consideration of current practice.

A considerable amount has been written about conservation philosophy in the nineteenth century and in particular the influence of John Ruskin and William Morris and the formation in 1887 of the Society for the Protection of Ancient Buildings (SPAB). The arguments that prevailed at the time were heavily influenced by the church and cathedral restoration being undertaken across the country. Restoration was seen as a destructive force and the imposition of the Gothic style on medieval structures described as vandalism. The church authorities were in total control of their own buildings and there was no regulatory system in place. The only influence that could be brought to bear was public pressure.

The first piece of legislation was the 1882 Ancient Monuments Protection Act which primarily identified a schedule of sixty-eight monuments in England, Scotland, Wales and Northern Ireland worthy of preservation. State purchase or guardianship of these monuments was possible with the agreement of the owner. The key figure in this was Sir John Lubbock, the Member of Parliament for Maidstone, who each year from 1873 to 1879 introduced his National Monuments Preservation Bill into the House of Commons, where it was repeatedly resisted on the basis of interference with owners' rights. The bill was described by a number of objectors as 'legalised robbery' although others argued that it did not go far enough. In the event, the 1882 bill as finally approved was a watered-down version of that submitted in the early days and it was to take another forty years for significant legislation to reach the statute book. It is interesting that Sir John Lubbock later purchased Avebury, Silbury Hill and West Kennett Longbarrow to save them from harm, thus demonstrating that ownership in the right hands provides the best form of protection. Although in 1882 the United Kingdom was a Christian country, only pagan monuments were protected. The church authorities successfully resisted the imposition of legislation at the time and for many decades to come.

Conservation interest groups such as SPAB have had a direct effect on the evolution of conservation legislation and their activities have often been inspired by particular events. The National Trust, formed in 1895, was originally established to protect open spaces in the Lake District but in 1896 purchased its first building, this being the fourteenth-century Clergy House at Alfriston, East Sussex, for £10. The National Trust would of course go on to become a major owner of historic buildings across the country. In 1897 the London County Council held a conference on Listing London's Historic Buildings and in 1898 obtained the power to acquire and preserve historic buildings.

The 1882 Act was superseded by the 1900 Ancient Monuments Protection Act. This allowed county councils to take direct action, and protection could be extended to medieval buildings, excluding ecclesiastical and occupied property. In his 1905 essay entitled 'The care of ancient monuments: a survey of methods of historic preservation in Europe', G. Baldwin Brown (Professor of Fine Art at Edinburgh) included a description of the legislation elsewhere in Europe illustrating how far Britain was behind on this issue. The only significant legal move towards long-term protection was the 1907 National Trust Act, which introduced the right of inalienable holding, thus setting the pattern for Trust ownership in perpetuity.

There were further Ancient Monuments Acts in 1910, 1913 and 1931 but little significant alteration to the legal protection of historic structures. In

1911 Tattershall Castle in Lincolnshire, a late medieval brick structure, was in a poor state of repair and the impressive Gothic fireplaces had been removed for sale to America, an event which caused public outcry. The castle was saved by being bought by Lord Curzon, who also rescued the fireplaces, once again demonstrating that acquisition was at the time the only effective method of control. During the debates on the 1913 Ancient Monuments Consolidation and Amendment Act Lord Curzon argued the need for protection of other types of building but was unsuccessful. The formation of the Council for the Protection of Rural England in 1926 was a protest against the effects of ribbon development and urban sprawl and, although not specifically dealing with conservation issues, was part of the process leading eventually towards planning control.

The term 'buildings of special architectural or historic interest' first appears in the 1932 Town and Country Planning Act. This Act gave powers to local authorities to make preservation orders for such buildings subject to the approval of the Minister who had to consider representations of those involved. Compensation provisions were included, and although demolition could be prevented, alteration or extension was uncontrolled. This was not an effective piece of legislation largely because no mechanism existed for identifying historic interest. Also in 1932 there was a plan by the Crown Estate Commissioners to demolish Carlton House Terrace, an event which again led to protest, as did the 1934 proposed demolition of Waterloo Bridge. The significance of this latter event was that it brought about coordinated opposition from the Royal Institute of British Architects, the Town Planning Institute, SPAB, the Royal Academy and others. The protest was unsuccessful and the London County Council went on to rebuild the bridge but the concept of coordinated protest was forged. Opposition to destruction emerged again in 1937 with the demolition of Robert Adam's Adelphi Terrace of 1768–72, and the protests led directly to the formation of the Georgian Group. It is worth remembering that this was the time when the Modern Movement in architecture was in the ascendancy and conservation was often represented as standing in the way of progress.

Widespread destruction by wartime bombing brought the plight of historic buildings to the attention of the public and Parliament. The concentration of attack on historic cities by the so-called Baedeker offensive resulted in destruction or serious damage to many historic structures. One reaction to this was the formation of the National Buildings Record (1941), the purpose of which was to record important historic structures with a view to their possible future repair. The damage brought about by the war also led directly to the 1944 Town and Country Planning Act which was primarily concerned with the problems of post-war reconstruction. Within this Act the Minister of Town and Country Planning was empowered to prepare lists of buildings of special architectural or historic interest for the guidance of local authorities. To deal with the issue of identifying historic interest, an advisory committee under the chairmanship of Lord Maclagan was established in 1945 to provide expert guidance. The subsequent recommended criteria for listing included a three-category grading system:

- Grade I – being buildings of such importance that their destruction should in no case be allowed
- Grade II – whose preservation is a matter of national interest and where destruction or alteration should not be undertaken without compelling reason
- Grade III – not special but of merit and worth trying to keep

The recommendations and criteria were approved. A Chief Inspector of Historic Buildings was appointed in 1946 together with a team of fourteen surveyors, later to be known as Investigators. Conservation legislation was strengthened in the 1947 Town and Country Planning Act, and it is noteworthy that the Minister was not simply empowered but given the duty to compile historic building lists. It was considered important that listing should be undertaken as a central activity and not delegated to local authorities where undue influence might be found, and at the time it was thought that maybe 100 000–200 000 buildings would eventually be listed.

In 1948 the government, in the form of the Chancellor of the Exchequer, established a committee under the chairmanship of Sir Ernest Gowers to consider the long-term problems associated with the preservation, maintenance and use of houses of outstanding historic or architectural interest which might otherwise not be preserved. This was again an issue brought about by the end of the war when a number of important historic houses were under threat. The terms of reference of the committee were not to consider whether such houses should be preserved but how this was to be done. The report, which runs to some eighty pages, is of interest in that it considered the roles of the various agencies, their powers and duties. The report identified the disparity between legislation relating to ancient monuments and that relating to historic buildings and the roles of central and local government. Included among the recommendations was a suggestion to amend the legislation to make it clear that historic building control should fall directly under the Town and Country Planning Acts rather than be referred back to the Ancient Monuments Acts. The committee recommendation regarding future activity was that a statutory body should be created for England and Wales and another for Scotland and entrusted with duties for furthering the preservation of outstanding historic buildings. Many, but not all, of the recommendations of the Gowers Committee were accepted by the government and were incorporated in the Historic Buildings and Monuments Act 1953.

The passing of the Act was followed by the establishment of the Historic Buildings Councils for England, Scotland and Wales. The primary function of these councils was the allocation of grant aid for outstanding buildings from central funds but their actions had a wider influence, especially in setting standards of repair. The Historic Buildings Council for England survived until the establishment of English Heritage in 1985.

In the 1950s the duties of the Minister of Housing and Local Government were primarily concerned with slum clearance and the accompanying redevelopment, these being the priorities of the day. In 1957 the then Minister,

Duncan Sandys, was in addition a key figure in the establishment of the Civic Trust, another voluntary organisation that would have significant influence. The 1950s also saw the establishment of the Victorian Society, again as a reaction to events. Demolitions in the early part of the twentieth century had led to protest movements and the formation of pressure groups, and there was a major outcry in 1962 regarding the proposed demolition of the Euston Arch. In spite of the protest the arch was demolished, but public opinion forced plans for the demolition of St Pancras Station and Hotel to be abandoned. At the time comprehensive development was in the ascendancy and it was only the premature release of illustrations of the proposed redevelopment of Piccadilly Circus that led to this scheme's failing to gain consent. The London County Council had been directly involved in the evolution of the project and it was left to public protest to provide opposition to the proposal.

The 1963 demolition of the Coal Exchange in the City of London and the 1965 demolition of part of Eldon Square in Newcastle were further contentious issues, and in the same year the Council for British Archaeology published its list of historic towns, there being 324 in England and Wales, 51 of which were identified as of national importance. Across the Channel in France legislation was enacted in 1962 to protect a specific historic area rather than individual buildings, and this almost certainly had a direct effect on subsequent legislation in the UK. The government activity at this time, led by Richard Crossman, Minister of Housing and Local Government, was the establishment in 1966 of its Policy Preservation Group that would in years to come significantly influence government attitudes. The 1967 Civic Trust-sponsored Civic Amenities Act, which was 'to make further provision for the protection and improvement of buildings of architectural and historic interest and of the character of areas of such interest', brought into the legislation the concept of conservation areas. Conservation legislation was further strengthened in the 1968 Town and Country Planning Act which included a requirement for listed building consent. Compulsory acquisition of listed buildings became possible, with minimum compensation payable if deliberate neglect could be proved. The 1968 collapse of the Ronan Point tower block gave further momentum to the reaction against comprehensive redevelopment schemes and fuelled the desire for conservation.

Although legislation provides the framework for control, it is policy interpretation and guidance that influence decision-making. Thus in 1968 the Ministry of Housing and Local Government decided to continue its promotion of conservation activity and commissioned four detailed studies. These covered Bath (by Colin Buchanan), Chester (Donald Insall), Chichester (G.S. Burrows, the County Planning Officer for West Sussex) and York (Lord Esher). The purpose of these studies was to identify and record the architectural and historic interest of the areas concerned and to suggest ways in which planning and conservation should be integrated. In 1970 the Department of the Environment (DoE) was established, linking planning and conservation more closely. The 1971 Town and Country Planning Act consolidated all previous legislation relating to listed buildings and conservation areas, excluding grants.

The 1970s were an active time; 1971 saw the publication of *New Life for Old Buildings* (DoE) and the Civic Trust/DoE promotional film, *A Future for the Past*. By 1972 around 1300 conservation areas had been designated and also in that year the DoE published *New Life for Historic Areas*. The 1972 Town and Country Planning (Amendment) Act brought in demolition control within conservation areas and the 1974 Town and Country Amenities Act introduced a duty on local authorities to consider further conservation area designation, including a duty to formulate and publish proposals for their preservation and enhancement. The battle between comprehensive redevelopment and conservation surfaced again, this time in Covent Garden where only a series of spot listings prevented extensive demolition.

Promotion of building conservation was given further impetus by the designation of 1975 as European Architectural Heritage Year. This was a campaign by the Council of Europe to encourage member states to see building and area conservation as a positive force for the future. In the UK the government proposed an initiative to provide low-cost loans to building preservation trusts which led directly to the formation of the Architectural Heritage Fund. This was also the year that saw the emergence of SAVE Britain's Heritage – a pressure group formed specifically to campaign on conservation issues. The 1976 Development Control Policy Note, which was a forerunner of the later Planning Policy Guidance Note, was an eight-paragraph document covering only two pages. Within a comparatively short time this had been superseded by DoE Circular 23/77, *Historic Buildings and Conservation Areas – Policy and Procedure*. This document consolidated and brought up to date central government advice; Part I contained guidance on policy and legislation relating to historic buildings, and Part II on that relating to conservation areas. There was guidance on the listed building consent procedure, repairs notices and financial matters, and within the document was a series of appendices including guidelines on alterations to listed buildings.

The first comprehensive survey of historic buildings started immediately after the war and continued slowly until the 1960s. By the mid-1970s resurveys had been undertaken in towns and cities but much of the country was covered only by the early lists which by then were seen to be inadequate. The lists were relatively short and excluded buildings that by the latter half of the twentieth century were becoming recognised as of considerable interest. Development proposals were often interrupted by spot listing and by conservation campaigns. In 1981 the demolition of the 1930s Firestone Factory on the Great West Road, just before listing, provoked Michael Heseltine, Secretary of State for the Environment, to instigate an accelerated three-year listing survey.

In 1979 the British Tourist Authority published the results of a study undertaken jointly with the Historic Buildings Council for England. This document, *Britain's Historic Buildings: A policy for their future use*, did not question whether such buildings should be retained. By then this was accepted policy and the study was intended to demonstrate how future use could be encouraged. The study concluded: 'Conservation must not be seen as a luxury, as an appendage to general planning policies, but as one of planning's central aims and tools.'

Major changes had been made to local government in the 1970s with the establishment of a two-tier planning system, and the Local Government Planning and Land Act was brought in to resolve the overlapping functions of local authorities. There was, however, no stated change in government views on historic buildings and the Secretary of State made it plain that the government remained determined to implement current policies to preserve the best of our heritage. The politics of the 1980s were much in favour of a reduction in government control and preferred to see work undertaken within the private sector or by quasi-autonomous non-governmental organisations, or quangos. It was therefore logical that the 1983 National Heritage Act should have included provision for the establishment of the Historic Buildings and Monuments Commission for England, later to be known as English Heritage.

If the question 'why conserve?' no longer needs to be asked, this does not solve the problem of what should be conserved. Are there, for example, certain parts of the environment that require a particular approach to be adopted and, if so, how should this be done? These questions and similar have been the subject of much discussion in the wider world and have over the years led to the agreement of various charters, conventions and recommendations drawn up by the international community. The purpose of these documents has been to establish general definitions and policies that could cross national boundaries and would encourage the recognition of cultural heritage. The relevance of this work can be seen later in the interpretation of legislation relating to building and area conservation and in the definitions needed to bring this about, one such document being the Charter for the Conservation of Historic Towns and Urban Areas adopted by the International Council on Monuments and Sites (ICOMOS) in 1987.

The 1973 DoE Circular *Conservation and Preservation* made the point that 'conservation of the character of cities should be the starting point for thought about the extent of redevelopment needs; and conservation of the character of cities should be the framework for planning both the scale and pace of urban change'. It is perhaps, therefore, surprising that subsequent legislation changes brought about a legislative separation between conservation and planning. The Planning (Listed Buildings and Conservation Areas) Act 1990, which remains the principal conservation legislation, is one of four 1990 Acts consolidating the legislation on town and country planning. The provisions in the Town and Country Planning Act 1971 regarding listed buildings and conservation areas with other related legislation are brought together in the 1990 Act, providing a virtually self-contained code for the protection of the architectural and historic heritage which tends to move away from the mainstream planning process.

Over the years it has been government policy to provide guidance on planning and similar matters and this is done in the form of departmental circulars. The supporting document for the 1990 Act is Planning Policy Guidance Note 15 (PPG15): *Planning and the Historic Environment*, published in 1994, which was also a consolidating document. PPG15 follows closely on the format and content of earlier circulars, including reference to policy issues and giving specific conservation guidance.

Responsibilities for conservation policy and casework fall to two different government ministries, the names of which frequently change. In the early 1990s some of the duties exercised by the then Department of the Environment became the responsibility of the Department for National Heritage, which itself was abolished in 1997 and the historic building functions passed to the Department for Culture, Media and Sport. The functions of the DoE, which became part of the Department of the Environment, Transport and the Regions, later the Department of Transport, Local Government and the Regions, and the then Office of the Deputy Prime Minister, are now exercised by the Department for Communities and Local Government.

The way in which the legislation regarding historic buildings and conservation areas has evolved over the last hundred years results in a complex series of controls backed up by detailed advice from both central and local government. It is true that owners of some sites will find themselves dealing with a range of controls which, although not in conflict, may nevertheless be confusing. The process of securing planning permission, listed building consent, scheduled monument consent or conservation area consent is frequently portrayed as the reason for delays and there is considerable pressure to ensure that the planning system delivers an acceptable performance for the development industry. In practice, it is not the legislation that is confusing; the difficulty is the degree of interpretation required and the variation in the abilities of those involved to recognise or accept the objectives.

PPG15, together with the accompanying PPG16: *Archaeology and Planning*, are currently under review and it is expected that these documents will be merged. Nevertheless, the majority of decisions regarding the care and conservation of the built heritage will remain with local authorities and it will be for these authorities to determine the relevant applications. In many cases applications are delayed or refused, not because there is a fundamental problem with a proposed change but simply because the applications are badly presented. It has to be recognised that particular aspects of the legislation have been brought in to deal with specific problems or as a reaction to events. Applications that follow the policies and guidelines of the planning authorities, including those of central government, have a much better chance of success, particularly if they are carefully considered and well presented.

The primary purpose of conservation legislation is the protection of historic buildings and areas relative to their significance. It follows that buildings listed Grade I come under the most scrutiny and that structures that are particularly rare will require special consideration. It is often forgotten that applicants for listed building consent are required to justify their proposals and need to show why works which would affect the character of a listed building are desirable or necessary. Applicants should provide the local planning authority with full information to enable the likely impact on the special architectural or historic character of the building and its setting to be assessed. The greater the understanding of the building, its character and significance, and the importance of its setting, the easier it is to argue for preservation or change.

The identification of buildings and areas for protection is simply the start of the process and recent studies and publications emphasise the need to recognise why these areas are important. We therefore see the promotion of detailed appraisals to identify and explain 'character' together with guidance on informed conservation. The historic environment is important in a number of ways and caring for this environment is a process that involves managing change. The purpose is to allow future generations to understand and enjoy those things that we value. This does not mean total preservation or retaining everything from the past but does involve informed judgements about value and significance. To be effective it is necessary to explain why a particular building or area is important and how it is best preserved. We cannot assume always to get the answer right and there is little doubt that future generations may establish alternative values. The policies and guidelines that are now available to us are based on a long-term development of conservation aspirations together with specific laws to protect those things which were seen to be of value at the time.

Changing attitudes to conservation are a subject that deserves detailed consideration especially where these attitudes are reflected in planning policies that have a direct impact on a changing world. Changes are now well documented: for example, the 2000 report *Power of Place* was a summary of the review carried out earlier in that year; the government's response to *Power of Place* was published the following year as *The Historic Environment: A force for our future.* In 2002 English Heritage and the Commission for Architecture and the Built Environment (CABE) published a report, *Building in Context: New development in historic areas*, which presents a view of what might today be considered appropriate new design in a historic setting.

At the beginning of this chapter reference was made to the present-day government view that the legislation had developed piecemeal. To address this, the Department for Culture, Media and Sport presented us in 2003 with *Protecting our Historic Environment: Making the system work better.* This consultation document sought views on updating and improving historic environment legislation to make it fit for purpose in the twenty-first century and included a number of suggestions for change. It was followed in 2004 by *Review of Heritage Protection: The way forward*, outlining the government's conclusions following the consultation responses received. Short-term changes within the present system and long-term measures involving primary legislation were proposed. The first of these short-term measures involving changes related to the listing process has now come into effect and more changes are expected to follow. In recent years successive governments have stated their support for conservation of the historic environment although among conservation practitioners and advocates the degree of that support will continue to be a matter of debate.

12 Conservation legislation in the United Kingdom: looking ahead

Colin Johns

The year 2004 brought a new approach to planning, replacing the development plan system which had been in place since 1968. The Planning and Compulsory Purchase Act 2004 introduced a new system comprising regional spatial strategies and local development frameworks, with the previous structure and local plans to be replaced over a four-year period. Further explanation of the government's approach to planning is contained within Planning Policy Statements and in particular PPS1 which defines and explains spatial planning: 'Spatial planning goes beyond traditional land use planning to bring together and integrate policies for the development and use of land with other policies and programmes which influence the nature of places and how they can function.'

At one end of the scale are regional spatial strategies for each region in England prepared by the relevant regional planning body, which will be the directly elected regional assembly if there is one. The next tier down is the local development framework comprising a collection of local development documents setting out the planning authority's policy. An essential requirement is for local planning authorities, essentially district and borough councils, to produce a statement of community involvement to explain to local communities and stakeholders how they will be involved in the planning process.

The structure of the system is set out in the 2004 Act, with further detail provided in the Town and Country Planning (Local Development) (England) Regulations 2004. Also relevant is PPS12: *Local Development Frameworks*. PPS12 explains that the local development framework should include a core strategy, site-specific allocations of land, and area action plans where necessary. The core strategy sets out the authority's spatial vision and strategic objectives for its area. The primary consideration appears to be an emphasis on planning for those areas that require intervention, which raises the question of how the historic environment will be recognised and protected within the new system.

The opportunity exists for the preparation of supplementary planning documents as part of the planning framework for an area although they do not form part of the statutory plan. Supplementary planning documents may be particularly relevant to conservation areas. As outlined in PPS12, 'Supplementary planning documents may take the form of design guides, area development briefs, master plan or issue based documents to supplement policies in a development plan document.' It seems likely that a supplementary planning document could set out an authority's policies for its conservation areas but the degree of detail may be less than existed in earlier plans.

For some time and in too many planning authorities, conservation officers have had inadequate influence on policy and control, or simply do not exist, leading to poor decision-making. The revised procedures do not address this problem but the changes taking place will in the long term transform the planning system. Viewed positively this could have the potential to remove the detachment between planning and conservation that emerged from the separation of functions described in the 1990 Town and Country Planning Act and the Planning (Listed Buildings and Conservation Areas) Act 1990.

Alongside these changes are other significant modifications arising from the *Review of Heritage Protection: The way forward*, published by the Department for Culture, Media and Sport in June 2004. The various studies undertaken and consultation papers produced in the early part of the twenty-first century were intended to address a degree of dissatisfaction among those concerned with the control of the historic environment. The government's aim in the review was to deliver a positive approach to managing the historic environment which would be 'transparent, inclusive, effective and sustainable and central to social, environmental and economic agendas at a local and community as well as national level'. In addition the aim was to provide 'an historic environment legislative framework that provided for the management and enabling of change rather than its prevention'.

The government's decision for the new approach to heritage protection is to look at short-term and long-term measures. The short-term measures are those introduced without the need for primary legislation. Longer-term measures will require primary legislation. As previously indicated, the basis for the new system is that it should be more transparent, flexible and comprehensible to all those who manage, own, live in and deal with designated sites on a day-to-day basis.

In the short term English Heritage took over the listed buildings system in April 2005 with administration transferred from the Department for Culture, Media and Sport. In a new move English Heritage is required to notify owners if an application or proposal to list their building is made. English Heritage must also consult local authorities on applications to list buildings. It has been recognised that owners and managers of historic buildings need good and precise information about what the listing of their property means, hence English Heritage will now include an information pack for owners giving detailed guidance about the implications of listing.

The present criteria for listing date back to the mid-twentieth century, although with periodic reviews and updating. A consultation document to eventually form the basis of new listing criteria was published in July 2005 and incorporates considerably more detail than was included previously.

141

The short-term changes are just the start of a comprehensive reform of the system for protecting and managing the historic environment. Proposals outlined in the *Review of Heritage Protection* include an integrated consent regime, unifying listed building consent and scheduled monument consent and administered by local authorities. It will also incorporate statutory management agreements, with reference to English Heritage as appropriate. Government will also consider further the findings of a research report on the possible unification of consent regimes, including the unification of planning permission and conservation area consent. This is all part of the longer-term package. In addition the government asked English Heritage to undertake a series of pilot projects to test the feasibility of sharing skills, expertise and good practice. Pilot projects are being undertaken in partnership with other government departments as well as local authorities, owner groups and the historic environment sector.

The unified *Register of Historic Sites and Buildings of England* is intended to bring together the current regimes of listing, scheduling and registration and will include World Heritage Sites. In addition there will be a local section which will contain a register of all conservation areas and other local designations such as local lists and registers. English Heritage will be responsible for compiling the main register and for national designations. Local authorities will be responsible for the local section, with local designations made against national criteria. The current Grade I and II* designations will be combined and renamed G1 and the current Grade II will become G2. The proposal that some buildings currently listed as Grade II might eventually be demoted to local lists was extremely unpopular in the consultation exercise and the government has indicated that it will not take this forward.

Entries in the current lists of Buildings of Special Architectural or Historic Interest are there to provide information but do not necessarily identify the special interest of a building. In many cases buildings were listed without being inspected internally and in these circumstances the list description could never be definitive. Under the new system there are to be Summaries of Importance. These are intended to be short, accessible and jargon-free, to enable whoever is using the document to understand what the designated item is, its physical and cultural context and its significance. To maintain credibility, Summaries of Importance will need to demonstrate consistency, which might require preparation of published selection criteria. Each new designation from April 2005 onwards contains a Summary of Importance.

When listing was first introduced it was decided that owners would not be consulted because this would allow them to alter or even demolish the building prior to the listing being confirmed. The new system suggests controls to ensure that this does not happen. The listing of buildings has been perceived by many as a closed process and there have been debates on whether or not the listing process should become open. The government believes that a more open system is necessary but that adequate short-term safeguards need to be put in place. Under the new system, from the point of notification to the owner and local authority of an application to designate there will be a period of public consultation, with interim protection incorporated.

Under the old system there was no formal right of appeal against a listing or scheduling at the point of designation, and in today's climate this is seen as undesirable. In addition to considering the rights of the individual, the government recognises that errors can be made and that there needs to be an appropriate mechanism to remedy these. There will therefore be a statutory right of appeal to the Secretary of State against the decision to designate or not to designate.

This new approach of enabling change has been taken up in the 2005 English Heritage Strategy. Here emphasis is given to a creative approach to conservation, endeavouring to protect the best of the past and allow it to be fully integrated in plans for the future. In order to do this it is essential that the value of buildings, areas and landscapes is properly recognised. Recent charters and similar documents such as conservation plans and conservation statements emphasise the starting point of identifying the key characteristics. Only when the importance is recognised can decisions be made about the acceptability of change.

The strategy echoes the government's objective of increasing access, educational opportunity and social inclusion in urban and rural areas as well as contributing to the national and regional economy. Attention is to be focused on buildings that are likely to fall out of use, especially public buildings such as schools, hospitals and town halls, and the Buildings at Risk Registers are to be linked with English Heritage grant schemes. Efforts will also be made to manage the countryside where significant changes are anticipated. These changes include agricultural restructuring, globalisation and the pressures for an increasing rural population. Adapting buildings to new uses is another theme of the strategy with special attention being given to churches, many of which are likely to be redundant in the foreseeable future.

In the present climate many conservation bodies view the government's commitment to the historic heritage with some dismay. This was made evident in submissions to the 2006 House of commons Culture, Media and Sport Select Committee's enquiry into Heritage Protection. A Select Committee can be a challenging forum and the committee found a number of weaknesses in the work of Department of Culture, Media and Sport. Within its fifty-seven recommendations the committee concluded that historic buildings were at risk unless the sector was properly resourced, and that the government should do more to provide English Heritage with the funding needed. This would be especially important if English Heritage was to fulfil the role expected in the heritage protection review. The lack of coordinated government thinking across departments was also viewed with concern. The government response to the report was unenthusiastic which further underlines the concerns expressed.

For the time being the Department for Culture, Media and Sport retains responsibility for listing and the work of English Heritage, and overall planning responsibilities remain with the Department for Communities and Local Government. This split responsibility is far from ideal and appears to be at odds with the requirement for more inclusive assessment and decision-making. Whereas in 1990 the Town and Country Planning Act and conservation legislation were separated, the 2004 Planning and Compulsory Purchase Act goes some way to integrate conservation as an essential component of the planning process. As long ago as 1968 recognition of the importance of building conservation was considered to be the starting point for assessing the scale and pace of change and there are good reasons to support this view.

There is no doubt that the public at large has a considerable affection for its historic environment in the wider sense. Maintaining and managing that environment involves a number of different organisations and agencies, not all of which see the conservation of the historic environment as their responsibility. This can lead to conflict, particularly where the work is undertaken by government or government agencies. The substantial growth of specialist groups and amenity societies from the middle to the end of the twentieth century is clear evidence of public interest and occasional protest. Comparisons have often been made between the effectiveness of the 'green' conservation lobby and of those seeking to protect historic buildings. The decision in 2002 of a range of heritage bodies and societies, including the National Trust, to form Heritage Link was a reflection of the need to influence the government and the public and private sectors.

Improving the environment and securing the reuse of buildings of historic value can make an important contribution to the regeneration of urban areas. In addition the regeneration of a single building or group of buildings and public spaces can initiate improvement of a wider area. In its 2004 report, *The Role of Historic Buildings in Urban Regeneration*, the House of Commons ODPM: Housing, Planning, Local Government and the Regions Committee concluded:

> The historic environment has an important part to play in regeneration schemes helping to create vibrant interesting areas, boosting local economies and restoring local confidence. When historic buildings, including churches and theatres, are no longer needed for their original use they are capable of conversion for a wide range of other purposes.

On the other hand, the vast majority of historic buildings are much appreciated by their owners who invest considerable funds in their long-term upkeep. Provided this appreciation of historic buildings is maintained, a substantial part of our heritage will be protected. The need is to focus attention on areas of change and to ensure that the protection of historic buildings remains centre stage. It will not be adequate to see this as the responsibility of central or local government, and communities and individuals will continue to need to take a leading role.

13 The role of the archaeologist

Peter Davenport

The legislation and regulations relating to historic buildings essentially have as their aim not so much the prevention of change as its management. It is true that such management may result in the refusal of proposals, but its underlying aim is to distinguish between inappropriate or undesirable alterations to historic buildings and appropriate and desirable ones. For this management to be successful, and reasonable in planning terms, it has to be based on historic, architectural and archaeological criteria that can be balanced against the demonstrated social or economic need for change.

Such criteria may be clear in very broad terms – for example, if a building is listed as of architectural or historic interest or is a prominent part of a conservation area, or perhaps is obviously an older part of its environs but not further differentiated or understood. But it is in the interests of both sides in the planning process to clarify exactly what in a building is valuable, or not, and why.

Once this is agreed, or at least laid out rationally for discussion, the particular proposals can be defended or amended on clear grounds and on a balance of needs. Ideally, of course, such an understanding of a historic building should be obtained as part of the design process, before or during discussion with the planning authorities. It is the role of the archaeologist in the planning process to obtain such an understanding.

It may surprise some to think of an archaeologist working above ground level, but in fact archaeologists have been investigating standing buildings since the late eighteenth century. Indeed, one of the original strands of archaeology was the detailed study of medieval churches in the earlier nineteenth century. Appositely enough, this was largely related to a church movement to reinstate older liturgical practices that involved sometimes substantial modifications in historic fabric to suit.

Historical – that is, documentary – research is an essential part of the study of old buildings. Building, or structural, archaeology, however, while never forgetting the contribution of documents to the understanding of a building, is the direct study of the building fabric itself. The building or structural archaeologist brings to this a blend of skills: logical deduction of sequence and relative date; art-historical dating from style and design features; structural understanding of how buildings of different periods are constructed; and academic knowledge, putting a building into a local, regional or national framework and making judgements on significance and importance.

Depending on the project, the archaeologist's role will in some cases be as a useful adjunct to, and in others as an essential member of, the design team, pointing out pitfalls to be avoided or opportunities to be taken. In all cases, fuller knowledge of a historic building will lead to a quicker and clearer resolution of conservation and preservation questions on the basis of agreed interpretation.

The archaeologist's role resides not just in the provision of knowledge and interpretation. It is also possible for him or her to act as a consultant to the developer – or indeed to the planning authority. There may be disagreements about the acceptability of proposals which the consultant archaeologist can help resolve, either by making an informed case for them or finding another way of achieving the design aim while satisfying the planning authority's objections. At an early stage, the archaeologist can also advise on the likelihood of proposals reaching determination without major objection, thus saving wasted time in redesign or appeal. Should appeals be necessary, a consultant archaeologist can act as an expert witness. As with all expert testimony, the archaeologist here appears as an unbiased expert, not for or against the client.

Within the planning process itself, local authorities by and large follow the guidance contained in PPG15 and PPG16 (DoE, 1990 and 1993), increasingly incorporated in local and regional plans. Sometimes, a specific supplementary planning guidance may be issued. Practically, apart from pre-application discussions, this means that a Local Planning Authority (LPA) will require information about the historical, architectural and archaeological aspects of the project as part of the application. This is usually in the form of an assessment report compiled by 'a suitably qualified person or organization'. LPAs usually provide a brief outlining the scope of the assessment – or in some cases English Heritage will provide the brief. This can vary from a preliminary assessment, including a site visit and archive scanning, map regression, a few photographs and a plan or two, to a much fuller study involving record and interpretation drawings, a photographic survey and an impact statement. This will largely depend on the original perceived importance of the building, but practice also varies from authority to authority.

If changes to the historic fabric are agreed and permission given, the authority may impose conditions. These usually require records to be made of before and after states, and/or demolition and rebuilding work to be monitored to record information revealed during the process. The detail and intensity of such arrangements will vary, but will always require the submission of a report on the observations made and recorded. There may also be conditions that refer to material used and design details. The archaeologist can contribute usefully to the discussions that will arise around these areas in relation to historic accuracy, likelihood and reasonableness.

The process of investigation will also apply when a historic building is to be restored, rather than significant alteration being intended as part of redevelopment, especially if it is part of a scheme to remove a building's later accretions or alterations to wind the clock back to a particular time.

Just because the intentions are academic rather than commercial does not mean that the works will not have to be properly justified. Even if the work does not come under any planning or protective legislation it still ought always to be done only on the basis of good information and understanding. Such an approach is often now termed 'Conservation-based Research and Analysis' (CoBRA). It might seem that this would always be the case in these circumstances, but in the past as much damage has been done to historic buildings by restorers as by developers. The philosophy behind all alteration works, from the manifesto of the Society for the Protection of Ancient Buildings in 1877 to the Venice Charter of ICOMOS of 1966, and underlying both modern practice and regulation, is first 'do no harm'; and if it is unavoidable, only do it on the basis of knowledge and understanding, and mitigate it.

There will be occasions when an owner simply wants to know as much as possible about the history and archaeology of a building for its own interest. No work to the fabric is envisaged, so there are neither regulatory nor philosophical aspects to be dealt with. In this case the archaeologist is often more relaxed, as the work is then entirely academic. Even so, the results must communicate clearly and without jargon. The following case studies illustrate some of the above procedures.

Acton Court, Iron Acton, South Gloucestershire is an example of a pure repair project. The house was bought at auction by a Historic Building Trust in 1984 but was quickly passed on to English Heritage when the importance of the building was properly realised. A programme of integrated archaeological excavation of the below-ground elements of the building, both under the standing structure and on the site of demolished ranges, and detailed study of the structure standing above ground confirmed the outstanding national importance of the building. This enabled designs to be prepared for conservation, restoration, further research and partial conversion for occupation. The brief was produced by English Heritage and the archaeologists provided a research design or method statement to meet it. Listed building and scheduled monument consent were granted.

The results were fascinating and important. The house had grown organically from beginnings as a thirteenth-century moated manor house, until by 1535 it occupied nearly all of the area within the moat. In that year the standing wing was thrown up hurriedly, but on a grand and luxurious scale, to accommodate the weekend visit of Henry VIII and Anne Boleyn. This wing formed the basis for the addition of further wings in the 1550s which surrounded the medieval core of the house. The moat was filled in and formal gardens laid out. Further additions were made in the late sixteenth century. Only the 1535 wing and part of the 1550 additions survived conversion to a farm in the late seventeenth century and further alterations down to 1980.

Study of the standing structure revealed not only its architectural interest and high status, but its extreme complexity as a monument. Detailed drawn records, assisted by photogrammetry, were made of external and internal elevations, cross sections, plans at different levels and details of mouldings and profiles. Materials used in the construction were recorded. These records were used to interpret and reconstruct, on paper, the different phases of construction and alteration to which the building had been subjected.

This information was then used to make decisions about how the building should be restored. For example, the 1535 wing had originally been rendered and limewashed white. Its present external appearance, however, owes more to alterations made in the 1550s and the seventeenth century. Restoration to the 1535 phase, while possible on the evidence available, would have meant the destruction of much of the 1550 work, as well as the later phases. The decision was therefore made to conserve as far as possible its present state externally. Repair, and even rebuilding as necessary – some parts were in terminal decay – was aimed at conserving features as they existed in the present fabric. There was to be no restoration to a certain phase or period. Internally, however, the archaeological study had made clear that the main interest of the building was its design as lodgings for the king. None of Henry's lodgings exist unaltered anywhere else. The only significant change had been the subdivision, vertically and horizontally, of the rooms in the early nineteenth century, partly using the Tudor panelling. Removal of these subdivisions and restoration of the panelling to its proper position, after full record and the granting of further listed building consent, allowed appreciation of the Tudor layout and the display of contemporary wall paintings of international value which had been discovered in the investigation phase. It also revealed further details, such as the royal garderobe. The 1550 wing, containing part of the long gallery, had lost its first-floor ceiling. The timbers above were revealed to have been recycled from an older monastic building, probably Kingswood Abbey, demolished at the Dissolution. It was decided, therefore, to leave the roof open so that it could be seen, a display decision at odds with historical accuracy, but at least based on a full understanding of the arguments.

It was also intended that the house should be returned to private ownership and work on making it habitable to modern standards also had to be accommodated. Much of this was on the ground floor, and new floors were unavoidable. All excavations for services and new floor foundations were carried out archaeologically and much new information on the pre-1535 house recovered before the new works, designed to be as unintrusive and non-destructive as possible, were undertaken. Although the house was listed Grade I and parts of the site scheduled as an ancient monument, the sale into private ownership was also hedged around with covenants to further protect this very important site. Every step in the research, restoration and conversion was informed by an understanding of the building predominantly derived from archaeological investigation. The work has now been fully published – see Rodwell and Bell (2004).

Acton Court provides an example of work being commissioned to a very high standard by a national heritage body on a building of national importance. A more typical example is work at the **Westgate Hotel**, Bath. This was a pub and hotel in the centre of the city occupying a narrow tenement with a Georgian facade (Figure 13.1). Proposals were brought forward by

Figure 13.1 The plain street elevation of 38 Westgate Street. Everything on this 'Georgian' facade except the masonry dates from 1969. Nonetheless, study showed a building with a complex building history.

the brewery to gut the building and turn it into a theme pub. The developer believed that only the narrow street facade was ancient and that he would have a largely free hand. However, the building was listed Grade II, and the local authority requested an assessment of its historical condition and value and the impact on them of the proposals. Historical and map research revealed that the pub had an uninterrupted history as a coaching inn and then as a commercial inn from the mid-seventeenth century, but that the oldest fabric probably related to a rebuild in 1797, with further modifications due to road widening in about 1810. The addition of a large Edwardian extension to the rear and one side had resulted in many alterations to the Georgian building and the whole complex was gutted and altered again in 1969. Nearly all the internal 'Georgian' decorative features proved to be of this date. Knowledge of the 1969 work was the basis for the developer's belief that nothing survived of the Georgian building.

The site visits carried out as part of the assessment revealed that, despite the gutting, much basic structure was still Georgian (and Edwardian). The assessment showed that, with careful design, much more of the Georgian fabric could be preserved than would be the case with the then current proposal, and that this would make consent much more likely. It also became clear that the building retained the potential for its fabric to answer questions about its history and that, for this reason, damage to it should be minimised. The building archaeologist then entered a series of discussions with the architects and relatively minor alterations were negotiated which achieved the design goals with much less removal of old fabric.

Permission for the revised scheme was granted with a condition that demolition and alterations be monitored and recorded. Access, unrestricted except by normal construction site rules, was granted to the building archaeologist to monitor and record all changes. Apart from clarifying and recording in detail the chronological and structural phasing of the building, these visits also led to the surprising discovery of almost a whole Georgian house carcase and parts of another embedded in the Edwardian extension (Figure 13.2). The inclusion of the architects' design drawings in the final report completed information on the current state of the building. The changes carried out in any successful project are part of the architectural history of the building.

Curiously, a planning application just a year earlier for a project in the same street, **33–35, Westgate Street**, Bath, also led to the 'discovery' of an unsuspected Georgian building and a redesign that not only preserved the old structure but probably resulted in a better design. A salutary point may be made in both these cases: if the building assessment had been commissioned *before* designs had been put together, much time and money could have been saved. Again, the LPA requested further information on the project as the footprint of the proposal included the curtilages of several listed buildings. The applicant believed that only the frontages were at all historic, as the rear and greater part of the block was occupied by a newspaper printing workshop that had been established in 1923. However, the assessment made it clear that the workshops had been built around the core of a purpose-built brewery of c.1809–10, all but the stables of which had survived conversion. A lease of the latter date had been drawn up shortly after construction and contained a description of the new building, which matched the four-storey central block.

Preliminary study on site confirmed the existence of much unusual Georgian fabric. Further study and a full analysis required by the LPA revealed that this was a structurally well-preserved brewery, with yard and front office and brewer's house, which was an extremely rare and unusual survival from this period, and of national significance. The building also revealed important historic links as in the 1860s, after the end of its use as a brewery, it had contained the offices of Isaac Pitman's press and Phonetic Institute, which promoted spelling reform and Pitman's shorthand. Major redesign this time enabled the conversion of the main block into flats and the

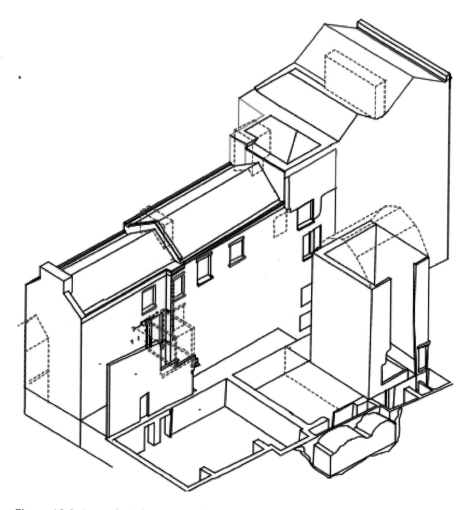

Figure 13.2 An analytical isometric drawing of 38 Westgate Street showing the surviving Georgian fabric after the notional removal of the later structures. The dashed lines show the lost Georgian features known from fabric evidence. Such drawings are a useful way of both demonstrating and recording historic building features.

creation of new build on the site of the 1920s and 1930s printing sheds. This required clever engineering and design solutions as the Georgian brewery was, shall we say, economically built. The frontage buildings were converted to shops and offices. Monitoring by the building archaeologist during breaking out and demolition works kept destruction to the agreed minimum and allowed recording of information as it was revealed. This project underlines the importance of obtaining the best information possible even on sites that do not at first seem important. The cost of the assessment report that first flagged up the problems was a very small proportion of the extra cost spent on revising designs.

A building that was known from the beginning to be of interest and importance was the medieval **St Ann's Chapel at Tal y Garn**, Pont y Clun, near Cardiff, a scheduled ancient monument (Figure 13.3). In 2004 this was

Figure 13.3 The chapel at Tal y Garn from the north-east. The seventeenth-century masonry is clearly recognisable as the darker stone on the left corner and the upper right corner. The tiny window and the wall around it are thirteenth or fourteenth century.

a roofless ruin, but until 1887 it was the parish church and until 1926 served as the church hall and Sunday school. The parish now hoped to restore the shell as part of new parish rooms. Cadw (the Welsh Assembly government's historic environment division) required a full analysis and record of the building, as well as an investigation of the churchyard.

While superficial study of the building made it clear that there was sub-stantial medieval fabric, and that the large windows in the south wall were probably the work of a late seventeenth-century benefactor, little was known of what was significant. A simple but accurate measured survey was made and into this were fitted rectified photographs to provide a detailed masonry record. Analytical drawings were prepared from this, both of the masonry and of the interior plaster, several coats of which survived (Figure 13.4).

From these studies it was possible to recognise that not only had more of the chapel been rebuilt in the late seventeenth century than was originally thought, but the roof had also been replaced at that time as well as in the later nineteenth century. The latter fact was evident from the fallen debris all around the site. It was also possible to show that the proposed door required in the north-west corner of the old building would be cut through seventeenth-century not medieval fabric which might be more acceptable to Cadw than if this were older. Nonetheless

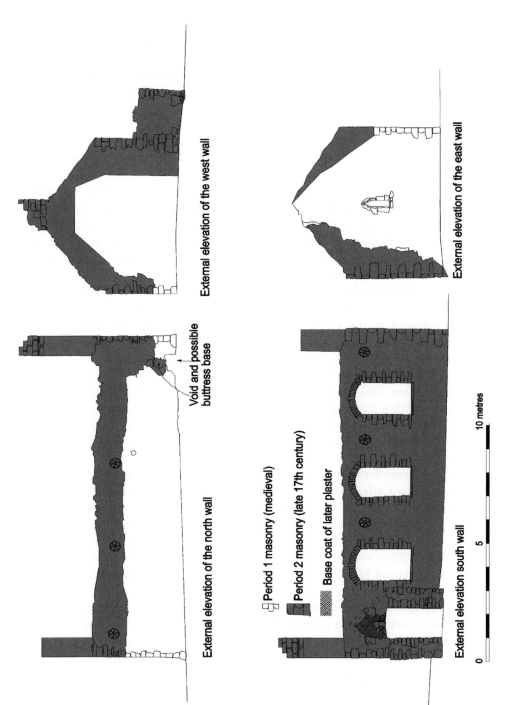

External elevation of the west wall

External elevation of the east wall

Void and possible
buttress base

External elevation of the north wall

▢ Period 1 masonry (medieval)

■ Period 2 masonry (late 17th century)

▨ Base coat of later plaster

External elevation south wall

0   5   10 metres

Figure 13.4 Analytical elevation drawings of the medieval chapel at Tal y Garn. Only the external drawings are reproduced here. The origi-
nals are in CAD format, allowing the production of working drawings if necessary. The project also included the production of an archive
of record photographs to complement the measured survey.

153

the study also confirmed the importance of the surviving medieval fabric. While the building was small, much investment had been put into importing high-quality oolitic limestone from England for quoins and the surviving window. The survey also showed that the building had undergone significant movement and needed 'beneficial reuse' if it was to be retained at all.

Study of individual buildings is of great value, but increasingly, the context of historic buildings needs to be addressed. This was the case in another development in Bath that was more about refurbishment than major alteration, involving a complex of interlocking properties of various dates including an early shopping arcade, **The Corridor**, built between 1827 and 1833 by Henry Edmund Goodridge. Fortunately, discussions allowed proposals to be made and tested against the historic constraints at an early stage and straightforward and mutually satisfactory solutions were found in nearly all cases. No major redesign was necessary. Part of the complex had been used in the 1880s as a photographic studio by a pioneer of cinematography, William Frieze-Green, and there was discussion as to whether it was appropriate to demolish a corridor that led to his studio and that appeared to be original. Detailed study made it clear that although the plan form was that of 1827–33, the actual fabric to be demolished was all post-1960. It was agreed that it could be removed but that the entrance to the passageway should be marked in the design of the reinstated shop front, itself based on original designs. The original and rear part of the passageway and associated staircase was retained beyond the area that needed demolition.

In relation to this development, another issue arose which is set to become more important in historic centres. This is *characterisation*. This concept is of particular importance where conservation areas are concerned. Many LPAs have not carried out characterisations for conservation areas or require more detailed assessment of parts of conservation areas or other areas for which historic or conservation issues are relevant. As this development affected most of a city block, within a conservation area and containing many listed buildings and surrounded by them, the LPA asked for a character appraisal of the development area and its environs to guide them in judging the individual aspects of the proposals. It was commissioned in this case from the buildings archaeologist, who in the capacity of architectural historian and urban archaeologist was able to bring the appropriate expertise. Often the project will require a small team with appropriate skills to work together. The issues to be addressed in such characterisations have been put together in guidance from English Heritage (English Heritage, 1993). The following resumé is quoted from Picard (1996):

A character appraisal of a conservation area or characterisation of part of one should include:

- the origins and development of the topographic framework of the area
- the archaeological significance and potential of the area, including any scheduled ancient monuments

- the architectural and historic quality, character and coherence of the buildings, both listed *and unlisted* [my emphasis]
- the character and hierarchy of spaces and townscape quality
- the prevalent and traditional building materials
- the contribution made by greens and green spaces, trees, hedges and other natural or cultivated features
- the prevailing or former uses within the area and their historic patronage, and the influence of these on plan form and building types
- the relationship between the built environment and landscape or open countryside, including the significance of particular landmarks, vistas and panoramas where appropriate.

Obviously, not all of these will be relevant in every case.

LPAs are increasingly requesting such characterisations and only rarely providing them themselves. Fortunately, however, they usually only arise where a project has a wide impact or is in a particularly sensitive site or area.

It has become apparent that the role of the buildings archaeologist is important, and in certain developments can be pivotal. Proceeding with design proposals in happy ignorance of their impact on historic fabric and environment is decreasingly a sensible option. In sensitive areas it can be expensive folly.

References and further reading

Department of the Environment, Planning and Policy Guidance Note 16: *Archaeology and Planning* (PPG16) (TSO, London, 1990).

Department of the Environment, Planning and Policy Guidance Note 15: *Planning and the Historic Environment* (PPG15) (TSO, London, 1993).

English Heritage, *Conservation Area Practice: English Heritage guidance on the management of Conservation Areas* (English Heritage, London, 1993).

ICOMOS, *The Venice Charter (1966)*, International Charter for the Conservation and Restoration of Monuments and Sites (Paris, 1966).

Picard, R.D., *Conservation in the Built Environment* (Longman, Harlow, 1996).

Rodwell, K. and Bell, R., *Acton Court: The evolution of an early Tudor courtier's house* (English Heritage, London, 2004).

14 Preparing the conservation plan

James Maitland Gard'ner

Introduction

The conservation plan is a document that identifies why a historic place or site is of value and formulates policies to assist in the retention of that significance in any future use, redevelopment, alteration or repair. It should inform, complement and, where appropriate, draw upon other documents such as historical studies, condition surveys, feasibility studies, and management and business plans.

Conservation planning was developed as a tool for balancing the different and often conflicting needs of conservation and development. Developed in response to *The Australia ICOMOS Charter for the Conservation of Places of Cultural Significance* (Burra Charter) – a document that puts understanding the significance of a site in a central position – the conservation plan can also serve as a tool for managing the conflicting values that various groups may place on a site. Used extensively in Australia and New Zealand, the conservation plan has gained currency in the United Kingdom since it was adopted by the Heritage Lottery Fund, who saw it as a mechanism to ensure that they were funding projects that did not harm the heritage assets they sought to protect. English Heritage and Historic Scotland adopted the conservation planning process as a method for better understanding their own properties in care as well as those they grant aid. Conservation planning also ties into the concepts of the local authority conservation area appraisals, Church of England conservation plans for cathedrals and statements of significance for parish churches, and the Countryside Commission's Quality of Life Capital.

All conservation plans must engage with those who have an interest in the heritage site and should be based upon a common intellectual process that covers the following concepts:

- understanding the site
- assessing its significance
- identifying how it may be vulnerable
- defining policies for its continued retention.

However, the conservation planning process should not be seen as strictly linear; issues identified within the policies, for instance, may raise questions that require further research and understanding of the site. Although often commissioned as a prerequisite of grant aid from bodies such as the Heritage Lottery Fund, the conservation plan should be used to assist in the day-to-day maintenance and management of the site as well as to inform proposals for change.

While the conservation plan establishes what is important about a site, why and to whom, a management plan will develop the policies and use them to inform a management regime or repair strategy, or to develop an action plan.

Commissioning a plan

The document itself may range from an outline plan of a few pages in the case of a conservation statement for a small individual building or monument, through to a published document running to several hundred pages with appendices in the case of a large complex site. It is important that the limitations on time and resources are recognised and are used to inform the briefing process to ensure that the final document provides the guidance required to inform the management of the site. Where in-house resources, skills and knowledge are being utilised in the preparation of the conservation plan, a realistic assessment should be made of limitations of time and expertise in order to identify when outside specialists should be engaged. As for any commissioned piece of work, a clear brief that sets out the requirements, scope and time frame of the project is necessary to ensure that costs can be accurately estimated and that the requirements of the conservation plan are fully communicated between those commissioning the plan and its authors.

Where the preparation of a conservation plan is being competitively tendered, a suitable balance of quality and cost criteria is needed to ensure that the final document is of sufficiently high quality as well as representing value for money. A useful alternative to traditional tendering is to request a project design within a set budget cost. The cost of researching, writing and publishing a full conservation plan will depend on the size and complexity of the site, the amount of information already available, the scope of the brief, the level of detail required and the amount of consultation required with interested groups. The preparation of a full conservation plan can add between £10 000 and £50 000 to the cost of the baseline surveys and research that might be required for the development of a capital works project or to facilitate effective site management. The conservation statement offers a cheaper alternative, albeit one more limited in scope and ultimate usefulness than the full plan.

The conservation plan should be coordinated by a single appropriate specialist, often an archaeologist, architect or other heritage professional. It is important that a single consultant take responsibility for the finished document, drawing together the various pieces of information or research to ensure that the document is written in a consistent style, is clear and meets the brief. It is seldom that a single person or even organisation will have the in-house skills to tackle every aspect of anything but the simplest of sites. Specialist disciplines required may include architectural history,

157

buildings analysis, metric and other survey, archaeological investigation, historical paint research, dendrochronology, technical materials research and ecological assessment, as well as facilitators to encourage meaningful community consultation. Precise briefing and coordination of other consultants is necessary to ensure that best use is made of limited budgets and that the information and analysis they provide meet the requirements of the overall document. Research should only be commissioned to meet the requirements of the conservation plan; research for research's sake will lead to wasted money and may not yield the information required.

Engaging stakeholders

The first stage of the conservation planning process – identifying and engaging with those who have an interest in a site (stakeholders) – is one of the most critical for ensuring the conservation plan's success. Although most often commissioned by the site owner or organisation charged with the care of a historic place, a successful plan should involve those directly responsible for the care and presentation of the site, those who use it, and the various statutory authorities and other interested groups. Establishing who should be involved in the drafting of the conservation plan and which groups or individuals should have the opportunity to comment will take some thought and may require brainstorming by the site owners and their consultants. The conservation planning process should allow for proper community consultation and for all views to be taken into account. It is only through identifying where the values of different groups or individual stakeholders conflict that effective policies can be developed for protecting the significance of those aspects of the site where there are competing interests, whether this potentially involves minor repair or major redevelopment.

Understanding the site

A full understanding of the site is a critical precursor to any conservation project. This section of the plan should provide an understanding of the site through its history to the present day. It will bring together basic information on the location, nature, ownership and current management of the site together with historical research. It should include a summary of all the different types of heritage that the site contains – architectural, archaeological, ecological, and so forth – and a history of its development from the earliest times through to the present day. Map regression is a useful tool for illustrating the series of changes that a historic site has undergone during its development. Any gaps in the history should be identified and further research conducted. As part of this process, historic and current uses of the site should be identified. Drawings, photographs and illustrations communicate this often complex information in an accessible and easily understandable way.

This section of the plan should not be burdened with an overly detailed historic study. For complex sites, an appended gazetteer that lists and describes individual buildings or elements, or room-by-room data sheets for historic interiors, will allow this information to be communicated without reducing the readability of the plan.

Assessing significance

The purpose of assessing the significance of a site is to identify attributes upon which we place value. The articulation of these values is necessary to allow informed decisions to be made about the management and development of the site. The values by which a site may be designated – 'special architectural and historic interest' in the case of listed buildings,[1] 'national importance' for scheduled ancient monuments[2] or 'special architectural and historic interest the character or appearance of which is worth preserving or enhancing' for Conservation Areas[3] – may only be a small part of why a place is significant to the wider community. As well as identifying any statutory and non-statutory designations that apply to the site, this section should also consider wider values including:

- architectural, aesthetic or natural beauty (design, artistic merit, craftsmanship or appearance)
- archaeological importance (the value of the historic fabric both above and below ground and what it can reveal of the development and use of the site)
- historic importance (associations with significant people or events)
- scientific values (technical innovation, ecological or geological)
- use (the value of the site owing to its historic or current function)
- community or social values (spiritual, commemorative, political or personal)
- artistic or literary associations (such as references to the site in painting, literature or film)
- public amenity values (including recreational use, views and open space)
- educational value (current educational value and potential for increased understanding).

The range of values a building, monument or site may have to different groups or individuals will differ and part of the purpose of this process is to identify and help reconcile these differences. To ascertain the full range of values that a site may hold, it will be necessary fully to engage with local people and other stakeholders. It may be necessary to facilitate dialogue through community consultation, especially in the case of complex sites or those with diverse stakeholder groups.

It can be useful to identify those elements of the historic place that are crucial to its significance and cannot be lost or compromised, and those that are of lesser or negative value. It is recommended that all aspects of significance should be articulated as paragraphs or bullet points rather than

159

given a numeric rating, as it is necessary to communicate why something is of significance and to whom in order to justify the assessment of value. If the importance of an aspect of the site is unknown, recommendations for further research or community consultation should be identified within the plan.

Identifying management issues and vulnerability

Having established the significance of a site, it is necessary to identify how aspects of the site may be at risk or vulnerable to change. There are many ways in which the site may be put at risk, including:

- lack of resources
- development proposals both on the site and on neighbouring sites
- physical condition or deterioration
- multiple ownership
- changing management structures or policies
- lack of understanding or knowledge about the site
- the current use or lack of appropriate uses
- actual or potential damage from site users or vandalism
- changes in conservation philosophy
- lack of traditional materials or loss of craft skills
- poor or inappropriate previous repair
- legislative and regulatory requirements (the provision of access for the disabled, for instance).

Conflicts in the values of the site or between different stakeholders need to be identified and their potential impact on the site assessed. A lateral approach to identifying vulnerability is necessary in order to consider the wide range of factors that have potential to place the significance of the site at risk. To establish these, dialogue with those responsible for the management of the site and neighbouring sites, as well as outside organisations such as the local planning authority, may be necessary. Again it is necessary to clearly articulate vulnerability in a way that links the identified risks to the historic fabric or specific aspects of the site.

Setting policies

The overall purpose of a conservation plan is to set out policies or guidelines that protect the site's significance and inform its ongoing management as well as future development. Policies should be developed that cover all aspects of the significance that have been identified as vulnerable, as well as providing guidelines for the day-to-day management of the site. These policies should be considered within the wider context of the legislation and planning policy relating to the heritage site, and may include the identification of new uses, the provision of disabled access, the establishment of an appropriate palette of materials or techniques for repair, the identification of appropriate sites for development or new facilities, the introduction of new services, the identification of repair priorities, the provision of security, guidelines for land management, and control of visitor access, car parking and servicing.

One way to set out policies is as a series of aims and objectives, specifically tailored to the management issues or proposals associated with the site. Where appropriate the policies should be tied to other more detailed organisational policies such as the access audit or disaster plan. For example, a typical policy from the Whitby Abbey conservation plan is as follows:[4]

Aim:

- To preserve the historic and ecological character of the landscape of the headland.

Objectives:

- The pattern of historically significant boundaries and walls should be retained and new alignments should not be introduced.
- Further deep ploughing of ridge and furrow should be avoided.
- Any conservation work to masonry should ensure that significant lichens are preserved.

A variation on this method is to link high-level policies with specific management issues and to identify appropriate options, as in the Lincoln Cathedral conservation plan:[5]

Policy:

- Signage and interpretation panels in the Cathedral and Close should be kept to the minimum necessary and any negative effects on the fabric and visual amenity minimised.

Management considerations and options:

- The use of logos on signs to convey information can be more discreet than off-putting written messages.
- Freestanding signs have the advantage of not damaging the fabric and are potentially less harmful to the visual amenity of an area.
- Continuing interpretation of the Cathedral and Close should be seen as an important element of their conservation and as an opportunity for contributing to lifelong learning.

Specific policies should be developed where there are known proposals for improved access, development, reuse or major repair of the historic building. The policies developed should establish a set of controls that enable future change on the site to be managed, but having said that, they should not be so overly prescriptive as to stifle imaginative proposals for the development or management of the site. Policies from the plan may need to be translated into specific documents such as the maintenance plan to inform the day-to-day care of the site.

Implementation

If full use is to be made of the conservation plan, its implementation needs to be considered. This will involve review and comment by those charged with the ongoing management of the site and other interested parties and site users. Without obtaining agreement with those responsible for the management, development and control of the site, a conservation plan is unlikely to be effectively implemented and will not be used to inform proposed work to the site. Statutory bodies may wish to implement management agreements with the site owner – for example, Heritage Management Plans or Countryside Stewardship Agreements – and the conservation plan can form the basis of these. The conservation plan for a larger, more complex site can also be adopted by the local planning authority as supplementary planning guidance, thereby giving the plan some statutory force in its own right.

Conservation plans should be used both to inform the development of new proposals and to review existing documentation for assisting the management of the site. These documents may include:

- maintenance and management plans
- further specialist research
- access audits and plans
- masterplanning
- development and repair proposals
- condition and quinquennial surveys
- historic garden restoration plans.

All management documents, policies and proposals should be tested against the conservation plan, and policies should be developed into more detailed guidance for those managing the site day to day.

Management regimes, the needs and desires of site owners and users, and the values the community may place on a site will inevitably change over time. To ensure that the policies and values contained within the conservation plan remain current it should to be reviewed regularly (approximately every five years) to take into account these changing circumstances, and to incorporate new knowledge gained about the site.

Publishing the conservation plan

The approved plan should be published widely, ensuring that the information and policies contained within it are made available to all those with responsibility for the management of or an interest in the heritage site. The document should be presented in a format that lends itself to easy storage and reproduction, and pages and sections should be numbered and clearly referenced to a table of contents. Hard copies of the plan should be supplied to the following people: the stakeholders involved in the preparation of the plan; site owners, staff and others who will be expected to use the plan to inform the management of the site; professional advisers working on proposals for the site (architects, landscape architects, etc.); funding agencies and other partners; and statutory bodies and amenity societies who have an interest in the site, including the local planning authority. A copy of the plan should also be deposited with the county records office and/or the National Monuments Record or equivalent, as appropriate. The copyright of any publication including a conservation plan normally resides with the author, although some corporate and government organisations may require that it be vested with the commissioning body.

The internet offers the opportunity for cost-effective broadcasting of the document to a wider audience including those with a more peripheral or casual interest in the site. The type of information included within the conservation plan needs to be considered so that the published document does not include either sensitive or confidential information. This supplementary information and extra detail is often better dealt with in appendices or a gazetteer.

Conservation statements

A conservation statement is essentially a shorter outline version of the conservation plan. It should follow the same format and rigorous thought processes, and contain maps, plans and photographs, but may be carried out in house utilising existing knowledge and research. Although they will vary in length and content, a conservation statement may be a shorter, more rapidly produced and less costly alternative to a full plan. A conservation statement may be used to provide guidance on small or non-complex sites, to inform the development proposals at an early stage before the full plan is ready, or as a preliminary stage in the preparation of a conservation plan. They are also a useful mechanism for engaging with stakeholders at an early stage, and can identify salient issues and potential areas of conflicting values, which can then be more thoroughly considered when preparing the brief for the full plan. The limited scope and lack of defined policies makes it dangerous to develop major redevelopment proposals on the basis of the conservation statement alone. Where an existing conservation statement is not able to provide adequate guidance on proposals affecting the site, a full conservation plan may be required.

Heritage impact assessment and mitigation strategies

While the policies contained within the conservation plan can directly inform the development of a maintenance plan for the day-to-day management of the site, carrying out a heritage impact assessment can be useful for testing development proposals against the policies contained within the plan. All new works, alterations to the historic fabric or changes to the management of the site have the potential to damage the significance of the historic place as well as to bring benefits. For the conservation plan to be effective in protecting the significance of the historic place, all proposals for change should be tested against the policies established and agreed within the plan.

The heritage impact assessment process should:

- establish the need for new work through a business case, access audit or the like
- state the anticipated benefits of the proposals to the heritage site itself as well as to those who manage or use the site
- identify which aspects of the site or elements of the fabric are affected and how the proposals may harm or put the significance of the asset at risk
- identify where further information is required prior to making a fully informed decision
- devise a mitigation strategy so that the impact on the significance of the heritage asset is minimised.

Mitigation can take a variety of forms, such as:

- undertaking further research or analysis to inform the design of the proposals and provide fuller understanding of the risks
- redesigning the proposals to avoid or minimise the effect on the asset
- not undertaking the work as a result of reassessing the need or providing the facility in another way
- selecting a different location, alternative materials or construction technique.

The heritage impact assessment and mitigation process can be summarised as follows:

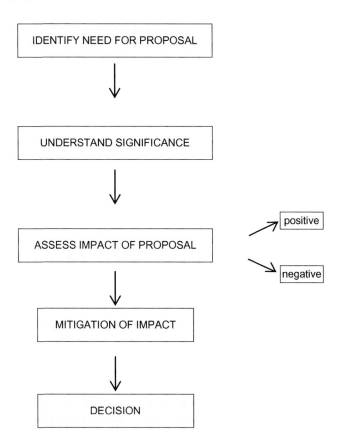

Table 14.1 Heritage impact assessment table.

Proposed work	Fabric affected	Significance	Impact/risks	Policy	Mitigation strategy
What work is proposed?	Which aspects of the site/ elements of fabric are affected by the works?	What is the significance of the elements as identified in the conservation plan?	What is the impact of the works on the significance?	Which policies within the conservation plan relate to this element/work?	What can be done to minimise damage to the asset, including further research?

The use of a heritage impact assessment table (Table 14.1) can provide an aid for formalising the decision-making process and communicate the justification for changes to the scheme or the way in which a site is used. The format of the table can be adapted to suit the needs of the project, and may include a column identifying where further information, research or analysis is requested. This column is especially important where proposals are being tested against a conservation statement rather than a full plan, as a lack of relevant policies or knowledge for informing the mitigation strategy will identify when a full plan may be required to provide the guidance needed.

Conclusion

The conservation plan is a powerful document that can provide the advice necessary for the management of a complex site and form the basis of other detailed documents such as maintenance and management plans. It should be used to inform new proposals and assess the impact of repair or redevelopment schemes. Conservation plans can also provide reassurance to funding and statutory bodies alike that redevelopment proposals do not damage or diminish the significance of the historic place. A conservation plan that has consulted widely and thoroughly considers the full range of issues associated with a heritage site is a valuable tool for reducing conflict between stakeholder groups.

Outline list of contents for a conservation plan

This outline list of contents should be adapted and built upon both for the processes of briefing and for writing the plan.[6]

Summary

A brief overview of the main conclusions of the plan.

Background to the plan

Including the reason for the plan, date of preparation, authorship, scope and any limitations on the study, the relationship to other relevant documents (e.g. business plan), and ownership of the plan (i.e. who is responsible for implementing it).

Stakeholders

Explain who has been involved in the plan and why. Describe the programme of stakeholder participation and the consultation process.

Understanding the site

Provide a general understanding of the site through time and as it is today. Describe the current management regime. Use numbering to organise complex information, and put detailed information in an appendix. This section should be well illustrated, and include a current survey plan showing existing features and facilities.

Note any gaps in knowledge, and whether they need to be filled through further survey or research in the short term or long term. More detailed information should be included in the gazetteer (see appendix).

Assessment of significance

Explain the different ways in which the asset is important and to whom. Note each of the statutory designations and what these tell us about significance. Provide a general statement of significance identifying each of the key values. Summarise what is important about each *chronological* phase of change in the development of the site. Identify what is important in each different *thematic* area: history, architecture, archaeology, art history, collections, library and archives, landscape history, ecology, geology, music, liturgy, community; describe the values people place on the site (users, local communities, tourists, employees). Identify features that are intrusive or detract from significance or have potential for development or change, including further access, use and enjoyment by a wider audience.

Put more detailed information about the significance of particular features in the gazetteer.

Management issues and vulnerability

Describe the issues facing the site and, in each case, how they could make significance vulnerable. Opportunities for enhancing the significance of the site can also be included. Consider the condition the asset is in and what issues this raises. Consider what the impact of previous conservation or repair work has been: are there lessons to be learnt? Comment on the appropriateness or otherwise of its current or proposed use. Describe known issues such as disabled access. Identify physical and other constraints affecting the site, such as lack of resources, statutory controls, and management policies. External factors (neighbouring developments, traffic, pollution) that may impact on the site should be identified and their effect considered. Note any gaps in knowledge or in the skills of the site management.

Policy aims and objectives

Identify how the site will be sustained or conserved. State the policy context (note existing legislation and statutory policies, as well as any other requirements or conditions) for the site. Include the overall vision for the site and the philosophy of conservation. General policies should include day-to-day management of the whole site, maintenance of historic fabric, the siting and design of any new development, reuse of historic buildings/spaces, the physical security of the site, visitor and vehicle access, interpretation, facilities and access provisions for the disabled.

Policies should be developed for all types of heritage associated with the site, such as archaeology, wildlife, buildings, collections, landscape, archives and technology. While the policies need to be specific to the site, they must also be consistent with local and national environmental or heritage policies, and with organisational objectives.

Appendices

Gazetteer

List each of the main elements, features, buildings, habitats or character areas for the site in a separate appendix. This ensures that detailed information can be found easily.

Photograph each element and provide a map to show its location. Use common sense to identify elements – these may be rooms in a building (and their contents), features of a landscape, areas of a site or individual structures. Ensure that the whole site is covered.

Supporting information

Provide copies or extracts of information essential to understanding the site, such as previous surveys (condition surveys, ecological surveys, metric survey), historical illustrations, and previous reports or research into the site. Information that is readily available elsewhere can simply be listed in the bibliography. Not all of this information may need to be published with the main body of the plan.

Management data

Include copies of designation documents, planning permissions, extracts from local authority planning policies and strategies and any agreements or conditions the site is subject to (including heritage, regeneration, open spaces or landscape policies). Identify the current pattern of use of the site, and its relationship to other adjoining heritage or recreational sites. Provide information about visitor numbers and the site's current use for employment, events, recreation and education.

Bibliography

Provide a list of sources relating to the site and its management, including previous surveys, published and unpublished reports, documentary sources and map. Cross-reference sources listed in the bibliography to the text of the plan where appropriate.

Further reading

Australia ICOMOS Inc., *The Australia ICOMOS charter for the conservation of places of cultural significance* (Burra Charter) (1979–99).

Cathedrals Fabric Commission for England and Association of English Cathedrals, Advisory Note 4, *Conservation Plans for Cathedrals* (London, 2002).

Clark, Kate (ed), *Conservation Plans in Action: Proceedings of the Oxford Conference*, (English Heritage, London, 1999).

Clark, Kate, *Informed Conservation: Understanding historic buildings and their landscapes for conservation* (English Heritage, London, 2001).

Heritage Lottery Fund, *Conservation Management Plans for Historic Places* (Heritage Lottery Fund, London, 2002).

Historic Scotland, Heritage Guide 1, *Conservation Plans: A guide to the preparation of conservation plans* (Edinburgh, 2000).

Semple Kerr, James, *The Conservation Plan: A guide to the preparation of conservation plans for places of European Cultural Significance*, 5th edition (National Trust of Australia (NSW), Sydney, 2001).

Endnotes

1   Planning (Listed Buildings and Conservation Areas) Act 1990 section 1(1).
2   Ancient Monuments and Archaeological Areas Act 1979 section 1(10).
3   Planning (Listed Buildings and Conservation Areas) Act 1990 section 69(1).
4   Whitby Abbey Headland Project Conservation by Kate Clark, June 1997.
5   From the Lincoln Cathedral and Close Conservation Plan by Dr Liv Gibbs, December 2001.
6   This is a summary of the list of contents included in *Conservation Management Plans for Historic Places* (Heritage Lottery Fund, 2002).

Appendix: North Hinksey conduit house conservation statement: consultation draft, May 2002

CONTENTS

1.0 Summary

North Hinksey conduit house is an important and intact piece of seventeenth-century industrial heritage that is surrounded by the suburban sprawl of modern Oxford. To ensure its continued survival regular maintenance of the structure and waterworks is required. Changes to adjacent farming practices and land use must also be considered in relation to the setting of the monument.

2.0 Background

This consultation draft of the conservation statement seeks to establish a first interim overview of the significance of the North Hinksey conduit house, and to propose draft management policies to assist in protecting vulnerable aspects of this site. It is anticipated that this draft statement will be reviewed and expanded over time by the stakeholders identified in section 7.0 of this statement.

This document has been prepared by Jim Gard'ner, English Heritage Historic Building Architect for South East Region, as part of an English Heritage-wide initiative. Existing documentary sources available through English Heritage have been utilised, and a site visit made on 20 February 2002 in the company of Dr Cockshoot, site key-holder.

3.0 Understanding the site

The North Hinksey conduit house is located adjacent to the A34 bypass road 4 km (2.5 miles) west of Oxford. The OS grid reference is SP49520505.

It is set on a hillside within grassed low-grade (class III) farmland with a twentieth-century housing estate towards the south-west, and earlier housing towards the village of North Hinksey. The adjacent farmland is in low-intensity uses such as horse-grazing.

The structure is a single storey limestone ashlar building measuring approximately 5.9 metres (19 ft 4 in) long by the same wide, and the walls stand up to 4.0 metres (13 ft) high at the gable ends. The walls are some 600 mm (2 ft) thick and the building measures 4.7 metres (15 ft 4 in) square internally. The side walls are divided into three bays by gabled buttresses. It has a pitched dressed stone slate roof. The front gable end is surmounted with heraldic arms and a two-light vent over the round-headed door. The door is painted softwood and has wrought iron ironmongery. The rear gable has a similar vent and heraldic device. Internally the building contains a large concrete water cistern that is still in use. The walls are covered in numerous examples of carved graffiti dating from the mid-seventeenth century to the present day, the majority of which is chiselled work of the eighteenth and nineteenth centuries. The route of the original conduit through the adjacent fields exists today as a water course.

The spring at Hinksey, called 'Reve Mores well', has been in use by the Blackfriars since before 1285 (Crossley, 1979). In the late twelfth or early thirteenth century the Botley family granted rights to the abbot to build a water house measuring 5.5 metres (18 ft) long by 4.0 metres (13 ft) wide over the spring in order to supply the Abbey of Osney (Page and Ditchfield, 1924). The current conduit house at North Hinksey was built in 1616–17 (Pevsner, 1966) by wealthy London lawyer Otho Nicholson as part of the Carfax Conduit. The conduit was built from 1615 to 1617 (Crossley, 1979) to supply water to the colleges and town of Oxford. Water from various springs was collected in a 9100 litre (2000 gallon) cistern within the conduit house before it was carried to Oxford in lead pipes encased in elm. The Carfax Conduit comprised two cisterns, the upper supplying the university colleges and the lower the town. As a result of road widening in 1787 Carfax Conduit was replaced with a new water house, and the elaborately decorated structure moved from central Oxford to Nuneham Park in 1789. The City Corporation purchased the water system in 1867, by which time it supplied little water to the city or university (Crossley, 1979).

The conduit house gradually became dilapidated and along with its immediate surroundings was placed in the guardianship of the Secretary of State on 2 July 1973. This was followed by substantial restoration works, specifically its reroofing and the reforming of the conduit cistern in concrete.

Car parking is limited to a small area of communal driveway adjacent to the neighbouring houses. Access to the public is via a farm track and the monument can be viewed from the outside only. A public path used to run from Oxford Castle to a ford adjacent to the site of the conduit house, but is no longer in use.

The condition of the monument is assessed periodically through visits by the English Heritage Technical Officer, and the cistern and pipework are inspected and cleaned six-monthly by a maintenance contractor. The

key-holder, Dr Cockshoot, reports six-monthly on the state of the monument, the site and its surroundings.

4.0   Significance of the North Hinksey conduit house

The North Hinksey conduit house is designated as being of national importance as a scheduled ancient monument (national monument number
28132), the scheduled area including a 2-metre (6 ft 6 in) boundary around
the structure. It is also a Grade II* listed building (UID 249742), although
on the list description it appears under the name 'Well House'.

The conduit house is the only standing element of the early seventeenth-
century water supply system for Oxford still on its original site, and marks
an important stage in the development of the city and its public services
infrastructure. Most other elements of the system have been relocated (in
the case of the Carfax Conduit), removed or built over.

The system as a whole, including this structure and the relocated Carfax
Conduit, illustrates a rare and important example of early private provision
of a civic clean water supply.

With the exception of the roof and cistern itself, the fabric of the 1616–17
structure survives in largely original condition and contains much archaeological evidence of its construction and use.

The carved heraldic stone cartouche over the door survives in readable
condition, providing a link with the conduit's builder, a figure of some
importance in the history of seventeenth-century Oxford. The graffiti, which
appears internally as well as on the exterior, provides a record of over 350
years of visitation, with the earliest sighted dating from 1634.

The conduit house is situated in an area of attractive semi-rural landscape
that forms part of Oxford City's green belt. Although only fleeting views
of the conduit house itself are possible from the A34 (Oxford Ring Road),
the site offers fine views over the 'dreaming spires' of Oxford. This view of
the city and university was immortalised in a number of paintings and
sketches by Joseph Mallord William Turner (1775–1851) in the late eighteenth century. Dr Cockshoot reported that the conduit house itself appears
in a painting by Turner (perhaps *Oxford from the South West*, c.1787–8).

The local community values this site as a place of interest within the
parish, and it is regularly visited by local history groups. At Heritage Open
Day weekends access to the interior is provided.

The scrub within the gully that follows the historic route of the conduit
pipe provides a wildlife habitat within the context of the farmland. The
gully may also contain archaeological evidence of the original conduit
pipe.

5.0   Defining issues (vulnerability)

The North Hinksey conduit house is a robustly constructed structure that
is inherently stable. However, deterioration of the fabric would occur over

time if regular maintenance of the structure and cistern itself were not kept up. The presence of a large volume of water within the well house makes it more vulnerable than many other structures to problems associated with moisture.

Vandalism poses a significant risk, given the monument's proximity to a major urban area. The long history of graffiti on the conduit house is testament to this. However, there is little evidence of late-twentieth-century vandalism owing in part to the relative remoteness of the monument from the road, and the fact that the historic route to Oxford Castle that ran near the site is no longer in use.

The setting and understanding of the conduit and the ecology within the gully are vulnerable to changes in farming practice such as the amalgamation of fields and the introduction of more intensive agricultural uses. Diseases such as foot and mouth have the potential both to limit access and to harm the monument through associated clean-up procedures.

Public access to the site is reliant on continued access through privately owned farmland. The current owner, Mr Carisbrook, proposes to demolish a farmhouse ('The Fold') adjacent to the entrance to the site and erect four new houses, the implications of which for the setting and access to the site are unknown.

New housing developments and roading improvements have the potential to adversely affect the setting of the monument and the views over the urban and rural landscape. This is evidenced by the visual as well as physical separation of the conduit house from Oxford by the A34.

There appear to be few health and safety issues associated with this site, although if access were to be provided to the interior on a regular basis measures would be required to protect the edge of the cistern. Requirements for improved access for people with disabilities in accordance with English Heritage policy and the Disability Discrimination Act 1995 have the potential to alter the setting of and access to the site.

6.0   Conservation policies

The ongoing maintenance of the structure, cistern and pipes, curtilage area and access should continue to be adequately resourced, with the regular maintenance programme continued and a system of quinquennial inspections instigated.

Repairs to the structure should be undertaken in matching materials and lime mortars using traditional techniques. Any repairs should be the minimum necessary and in accordance with English Heritage best practice standards. All repair works should be recorded and the information deposited with the NMR in Swindon.

The key-holder and others should monitor vandalism as part of the regular inspection of the site. Anti-graffiti coatings should not be applied to the stone fabric, and any cleaning of the structure should only be undertaken on the advice of recognised experts in the field.

Prior to any works or changes of farming practice that affect the former route of the conduit pipe, an ecological and archaeological evaluation of the gully area should be undertaken.

English Heritage should enter into dialogue with adjacent landowners, leaseholders, the local community and the local planning authorities to establish lines of communication in order to be kept informed of changes to land management regimes and proposed developments. Any proposed development or change to farm management should be assessed as to its effect on the wider setting of the monument.

Public access should continue to be limited to the exterior of the conduit house only. Access to the interior on Heritage Open Days should be supervised by English Heritage staff or key-holders.

A disability access audit was undertaken by English Heritage in June 2001. As access is limited to the exterior of the conduit house, the proposed alterations to the gates, paths and interpretation are likely to have only minimal effect on the setting of the monument and none on the conduit house itself. The design and siting of new interpretation panels and alterations to gates should be considered in order to minimise visual impact and potential damage to below-ground archaeology.

7.0   Implementation and review

This document is an initial consultation draft and should be reviewed by the key partners for their input and formal agreement. These include but are not necessarily limited to:

- English Heritage
- neighbouring landowners/leaseholders
- site users and residents of North Hinksey
- Vale of White Horse District Council (local planning authority)
- Oxfordshire County Council.

Once agreed, this conservation statement should be reviewed no later than May 2007, or earlier as need dictates. Unless proposals are developed that are likely to adversely affect the setting or fabric of the North Hinksey conduit house, it is unlikely that this document will need to be developed into a full conservation plan.

8.0   Appendices

8.1   Consultation

Dr Cockshoot, local key-holder
Fred Powell, English Heritage Technical Officer
Ray Phillips, English Heritage Project Coordinator

8.2   References

Cole, C. 'Carfax Conduit', in *Oxoniensia* (Oxford Architectural & Historical Society, Oxford, 1964), pp. 142–66.

Crossley, A. (ed.) *The Victoria History of the County of Oxfordshire: Volume IV* (Oxford University Press, Oxford, 1979), pp. 354–5.

Department of the Environment, List Description (1966).

Department of National Heritage, Schedule Entry (1996).

Lord Montagu of Beaulieu, *English Heritage* (Queen Anne Press, London, 1987), p. 92.

Page, W. and Ditchfield, P.H. (eds) *The Victoria History of the County of Berkshire: Volume 4* (St Catherine Press, London, 1924), p. 405.

Pevsner, N., *The Buildings of England: Berkshire* (Penguin, Harmondsworth, 1966), p. 186.

(The original document contained in addition a site plan, floor plan and photographs as Appendices 8.3, 8.4 and 8.5.)

15 Costing and contracts for historic buildings

Adrian Stenning and Geoff Evans

Introduction

Historic buildings have many types of owner – private individuals (accounting for some two-thirds of listed properties), local or national preservation trusts, commercial institutions and the various church organisations. Few clients have bottomless pockets; repairs to listed historic buildings cost, however, an average of 4.7 times more than those to unprotected – unlisted – structures.[1] The conservation of a historic building can cost substantially more than building a new structure of equivalent size and even a conservative assessment suggests that:

> Building owners or developers weighing up the relative merits of refurbishment as opposed to new build, may be surprised to discover that renovation projects can exceed 80% of the cost of a similar facility from scratch . . . the ratio is often around two thirds.[2]

Conservation projects also take longer; if a £1 million new building were to take six months, a £1 million repair contract might take a year or more. Cost prediction and control can therefore have a significant part to play in a historic building project, and a quantity surveyor with the necessary experience, or a professional possessing the same skill base, can be an extremely important member of the project team.[3]

So why are historic-building projects so expensive? There are several principal factors, each with cost implications, between the initial stages of a scheme and its conclusion. These can be identified as involving (1) risk, (2) complexity and quality and (3) approving and grant-aiding bodies.

Risk

For greenfield developments, the design and building work can be totally predetermined and it should be possible to run from start to finish with no variation. This is not the case when working with historic buildings where there is substantial risk of unknown and unforeseen circumstances that can easily compromise an initial cost estimate and adversely affect the final cost. It is rare that the building fabric can be opened up sufficiently to fully establish its condition – sometimes it cannot be opened up at all. No one can accurately predict what lurks behind panelling, beneath floor voids, in roof spaces (including bats and swallows), and so forth.

When works are under way, it is not only the unanticipated problem itself that may affect cost; the time involved in arriving at an appropriate solution may also be significant. Solutions are not always easy, and consultation with other professionals may be necessary. There may also be a need to gain the approval of the authorities and grant-aiding bodies. If the delay is significant, this will invariably lead to a financial claim by the contractor for 'lost' time. To allow for risks to be encountered and accommodated without compromise to the overall budget it is advisable to fix the contract period beyond what the contractor may initially feel necessary. Ample time also allows for samples of materials and workmanship to be adequately considered.

It is particularly at the stage of estimating and the production of schedules or bills of quantities – and, for the contractor, at the tendering stage – that conservation building projects require a different approach from new-build projects. This is the time of greatest uncertainty and therefore highest risk to accurate costing, as invariably at this stage only partial exposure, at most, of possible defects and failures can be made, and of course their remedies can only be ascertained once they are exposed. The timescale may not allow full investigation, the building may be in occupation, or the building may not yet be owned by the client. Moreover, investigative work at this stage may force a commitment to continuing with the works.

The architect and surveyor must make calculated judgements of potential problems and their remedies based on experience and on such careful site inspection as can be made. This may be limited to lifting floorboards, examining roof spaces, inspecting a church spire with binoculars, or assessing from the spring in the lead parapet gutters how much of the wall plate is rotted and whether the rot may be confined to isolated beam ends or whether a full flitch is required. The structural engineer's report, meanwhile, can have significant cost implications.

The naive cost adviser will only include those items the site inspection has exposed, and the estimate or bill of quantities will be insufficient: for example, costing for re-roofing but omitting to allow for the inevitable consequent repairs to decayed timber. The adviser will then struggle throughout the project, become unpopular with the client and other team members, and receive further appointments. The over-cautious adviser, on the other hand, will include not only those items the site inspection has exposed but also all the possible hidden problems, and the over-provision of costs will probably kill the project, or at least provide a licence for an astute contractor to quote high daywork rates. An experienced quantity surveyor will include items the site inspection has exposed, and in addition those items that are likely to occur but may at this stage be hidden (the calculated risk). This will give the project estimate a slightly over-provided appearance at the outset but the excess margin will usually prove to be required by the end of the project.

Complexity and quality

In greenfield development, standard descriptions and reference to codes of practice and British Standards are generally adequate. Works to historic buildings, however, generally present a broad range of difficulties. Workmanship requirements are often not evident from standard descriptions, while official standards may not apply to traditional materials and deviation from these may be required. The works are often site specific, needing to be sympathetic to their surroundings; access may present difficulties. There are sometimes timing constraints, as certain traditional building materials must be applied only at certain times of the year. Historic buildings have innate significance either because of their endurance through time or because of their architectural quality. A vernacular timber-framed barn and the most well-mannered classical edifice each have qualities that defy standard terms and descriptions. Inadequate initial understanding of a building can lead to inaccurate estimating and inadequate documentation that in turn will lead to potential problems later on site.

Any given level of complexity and required quality will have a lowest cost which cannot be undercut without compromise to quality through inferior workmanship or the use of inadequate materials. An underfunded project has one of three consequences: either the quality is compromised and the project 'fails'; or the contractor is made to take up the loss, a situation which, if repeated, will eventually cause the contractor to fail and therefore undoubtedly the quality will fail also; or, perhaps most likely, the contractor will make a claim, causing conflict, difficult working arrangements and possibly prolongation of the contract. To avoid such situations, and to achieve a competitive and accurate cost for quality work, it is important to provide correct advice in terms of procurement, schedule documentation and contract, and to set an appropriate contract period to allow for unknown factors and the preparation of sample materials and workmanship.

Approving and grant-aiding bodies

The restrictions, conditions and approved methods of working that a local authority may apply, perhaps in consultation with English Heritage, often have a cost implication. The perception is that grant aid from local authorities, English Heritage or the National Heritage Memorial Fund can increase the cost of a project as the outside bodies may impose their own solutions and standards that are beyond those the client might have wished for.

For many projects, however, it is essential to achieve maximum grant aid in order for them to proceed. Such aid as might be available is usually calculated early in a project's life, and it is essential that all the potential problems associated with the building are identified before this is sought as grant providers are seldom able to increase their original allocations. The quantity surveyor's presentation of the financial documentation can also make the difference between success and failure. It should be in a form that will withstand scrutiny and both maximise any grant and provide it in the shortest time – minimising the owner's outlay.

In short, a historic building project is likely to contain expensive items with a high level of uncertainty and complexity and to be dependent upon outside funding. An experienced quantity surveyor will be able to suggest options or alternative approaches either to a specific item of repair or to the overall approach to the entire project.

The brief and the budget

The required budget for a repair project will depend heavily on what the client aims to achieve, whether it is the National Trust commissioning a museum or a home owner mainly concerned with installing basic comforts such as heating and bathrooms. The historic importance of the building will dictate the quality or type of repair and therefore the cost. For example, there are various options in reinstating a missing area of ceiling plaster: cheapest is plasterboard and skim; lime plaster and galvanised expanded metal lath might cost three times as much; lime plaster and stainless steel expanded metal lath, five times; lime plaster and riven oak lath, seven times. Equally, with repairs to an oak timber frame it would be cheaper to renew in complete lengths rather than replacing timber by cutting out and piecing in or forming scarf joints.

Another consideration is the anticipated length of time until the next major expenditure. For masonry repairs to a provincial cathedral, such as Lichfield, it might be reasonable to assume an eighty-year interval before the next works to the spire (where scaffolding is a very substantial part of the cost) and therefore to carry out inexpensive but short-term mortar repairs would be questionable. Repairs to the tower might reasonably be anticipated at fifty-year intervals, and to the lower levels at twenty-five year intervals.

The likely programme of works also influences cost – for example, if there are to be delays while archaeological investigation takes place – although an initial opening-up contract may be advantageous in supplying information for the main contract. Another factor is the conditions that the conservation officer or English Heritage inspector may impose. A farm building at Harnhill, Gloucestershire, had one roof slope with natural stone tiles and one in corrugated iron, so could have been re-roofed as existing or entirely with natural stone; in such a case the authorities will surely have a view. At Truro Cathedral English Heritage required the pinnacles to be repaired with Bath stone, as existing, together with mortar repairs, while they could have been repaired using less expensive but equally performing French stone.

Contract documentation

For new build, the quantity surveyor's bible is the *Standard Method of Measurement*.[4] Conservation work, however, requires an item-specific approach, with the specification sufficiently detailed to avoid incorrect removals, bad or inadequate workmanship and the use of wrong or inadequate materials. Least of all is it possible to estimate on a cost-per-square-metre basis. To ensure the contractor has a full understanding of the works involved, especially if they are of a sensitive nature, the descriptions should include the location of the work, the materials to be used, the fixing method and fixing materials, the precise methods and timing, and the quality of finish. Some of these items may of course be covered by standard measurement clauses but others benefit from a full description; for example, in the case of skirting, in addition to stating that it should be fitted to match the existing, the description should state whether it requires – or does not require – packing to the wall. Either way, there is a cost implication to the contractor if such items are not mentioned. Bills of quantities for such work may be based upon *Standard Measurement* rules but expanded and amended to suit each individual area of the work to be undertaken:

> Practitioners involved in conservation will find that there is a need to introduce particular clauses of their own for inclusion in the project specification; over a period of time, a series of suitable specification clauses will accumulate which are then available on a needs basis.[5]

The requirement for such clauses will vary according to the sensitivity of the project, but without them underpricing is inevitable and will cause a difficult working relationship to develop, and may possibly result in compromised quality of workmanship and materials.

It is essential to assess the possibilities of hidden defects and to make allowances for provisional work, without which the project will invariably run over budget. At the budget stage it is reasonable to assume that 30–70% of the project cost will arise from areas that are not visible. These provisional sums will provide rates for use if unforeseen work is necessary and allow the contractor to develop a programme that will accommodate such work.

It is not good practice to place all risks on the contractor, as in a clause such as 'reslate the roof reusing existing slates supplemented with new matching slate to make up deficiencies'. The contractor may not have time to assess the situation and small contractors, particularly, who cannot afford to risk bankruptcy, will submit an inflated tender, while the clause leaves open the question as to what material is reusable.

The tender document is also a valuable post-contract **cost-monitoring tool**. With historic buildings as much as 50% of specified work might be varied as the fabric of the building is opened up and as scaffolding affords close inspection. The contract documentation should be written so as to allow for variation. Rather than specifying, 'carry out all masonry stone replacement shown on the drawing', which will be priced as a lump sum, items should be separately identified so as to provide options as the works proceed. Nor should too much be left to the tenderer's discretion: for repairs to a church spire, a clause that reads 'rake out and repoint all defective mortar' will probably lead to a post-contract dispute over the extent of repairs that were necessary. There should also be a definition of what constitutes 'defective', such as 'all cement mortar'.

Each repair should be broken down into separate operations and insist on item-by-item pricing. For example:

Carefully lift the cover flashings.
Take up the lead parapet gutter.
Renew the gutter boardings and bearers in treated softwood.
Reline the gutters in Code 7 lead.
Redress the cover flashings.

Indicating the location of each operation, proceeding room by room, roof by roof and elevation by elevation, has several advantages. It helps the contractor and makes it easy to compare specified items in the document with work on site, and to identify and price variations.

Rather than relying entirely on provisional sums or contingencies, the tender document should require **provisional quantities** – rates for work such as repointing, and roof timber and plaster repairs. Items that would have been daywork at indeterminate or arguable costs can then be costed at competitive rates as soon as they come to light. A project that includes provisional quantities is also more likely to be favoured by grant-aiding bodies. For indeterminate items such as door repairs, the tendering contractor should in addition be required to provide competitive **provisional hourly rates**, trade by trade, rather than provisional lump sums. Specified items should also allude to **quality** in order to minimise disputes and condemned work.

It is important to state the **contract period**. Allowing the tenderer to state the contract period usually results in an allowance for too little time and leaves open the likelihood of claims for extension of time. As the fabric of the historic building is opened up the unexpected will be encountered, and the thought and detailing required is inevitably counter to a very tight programme. Tendering contractors' offers to complete the works in a shorter period should be ignored. Historic building repairs require time to execute; the carbonation of lime or the splicing in of a piece of oak cannot be hurried.

Finally, the contract documentation should be **client specific** and specify times to avoid building work: for churches during funerals and Holy Communion, and on Sundays; early mornings if there are neighbours; examination periods if near a school, and so on.

Contract procurement

There are several types of contract that can apply to conservation work. The surest way to achieve quality and the best working relationship is through a **negotiated contract** with a specific contractor, who then becomes part of the team. The number of recommendable contractors for this route is diminishing, however, while many clients, such as government bodies, will not accept such methods. Although competitive rates can be achieved by a negotiated contract, these cannot be proven without the unethical practice of seeking tenders from contractors who will not be awarded the project. Also, slightly enhanced rates which may be acceptable can appear unpalatable, although – and again, it cannot be proven – it is quite likely that the final cost will be less as fewer confrontations and consequent claims are likely to arise. A negotiated contract is suitable for projects of any size, but especially small ones.

The surest way to achieve the lowest initial cost is through **single-stage competitive fixed-price tendering**, and to many clients it is the only acceptable route. This method, when combined with the use of inflexible standard methods of measurement, often results in claims by the contractor for extra work, a likely overspend and a more difficult working relationship. When used with good documentation, however, this method can achieve good results. It can be used for any size of project, but especially those of medium size where individual specialist packages are not necessarily large enough for subcontracted specialists and where the contractor's own craftsmen are able to achieve the results; very specialist works, meanwhile, can still be sublet.

A **two-stage tender** is also possible, with a first competitive stage and a second negotiated stage. An initial competitive tender is held based upon almost completed schedules or bills of quantities, while the second stage of negotiation provides the opportunity to supply further information to ensure that the work is fully understood and costed by the contractor, and to enable the costing of variations to be achieved with little or no dispute. This process works much as for single-stage competitive tendering but with the potential to make final adjustments, offering flexibility and ensuring the smooth running of the project and the resulting quality. This system, which may be used on any size of project, may still not be acceptable to some clients although it can mitigate against overspending.

In **management contracting,** the contractor takes on the role of site manager and sublets packages of work to specialist subcontractors, often pricing in competition. This can ensure that the works are carried out by appropriate specialists while the administration and programming can be efficiently carried out solely by the management contractor, who charges a percentage for overheads and profit. This method can be problematic if one subcontractor holds up another, with subsequent knock-on effects, and it is the management contractor's responsibility to sort out such situations. Management contracting can reduce the need for professional consultants but requires a level of trust as there is particular dependence on the contractor, and documentation has to be of high quality. This route is mainly suitable for large projects where substantial or highly specialised works are required.

Daywork ('cost plus') contracts are ideal for keeping firm control over building works in certain situations, but the disadvantage is the lack of cost control. Costs can be controlled or at least monitored by setting target times but this detracts from the initial benefit of working to day rates. It may be appropriate mainly on small projects.

181

The **guaranteed maximum price contract**, achievable in one or two stages, is unlikely to have the most economic initial cost but may provide some certainty of outcome. A minimum of 80% of the project should be fully designed before this type of contract is effective and it is therefore applicable to relatively few conservation projects.

Finally, **directly employed labour** can prove economical for organisations such as cathedral works departments, but all costs should be accounted for, including workshop costs, supervision and holidays with pay. The potential disadvantage of this method is loss of financial drive and the discipline to keep to the programme. English Heritage may grant-aid such work on the basis of a notional contractor's price.

Contract and contractor selection

The standard forms of contract, such as the JCT group of contracts, are appropriate for conservation projects, and the choice of contract should be made in relation to the total cost or duration.[6] Conservation projects are inherently expensive and often have longer site periods than new-build work but, although invariably demanding high-quality work, they are in essence usually quite simple in terms of the work being undertaken. The contract needs only to be adequate to administer what are largely traditional craft techniques. Also, quite often the level of opening up still required at the contract stage would make some terms of larger, more complex forms of contract difficult to meet, as all the information will not be available to the contractor. The simplest contract that protects the client and provides adequate insurance and instruction methods is to be recommended.

As for selection of the general contractor, with all the considerations of size and skill base, this will no doubt be based on interview and one's own past experience. In general, a contractor should be of a size suited to the project, with proven experience in the given field and with good management skills so that realistic contract programme times are set up with the specialist trades such as lime plaster. The contractor should preferably have in-house craftsmen in various trades – leadworker, plasterer, bricklayer, joiner, roofer, and so on – which avoids problems of zealous subcontractors submitting claims – but such contractors are increasingly difficult to find. Ideally, the contractor should have an apprenticeship scheme, as this indicates commitment, but this is becoming even rarer. Where highly specialised work is required or when there is an area that the contractor does not cover well, specialists should be brought in – for example, for ornamental plastering, mosaic conservation or wall-painting conservation. If the management contracting route is to be taken, particularly good management skills would naturally be required as well as experience in the field so that realistic contracts are set up with the various specialists.

Contractors who appreciate the complexities of repairing historic buildings will price the specification adequately and programme the works appropriately. In general a contractor without experience of repair contracts will underprice, and not understand the time requirements for conservation work and the delays as defects are discovered and repair techniques formulated. The expectation to make a quick profit results in hurried workmanship. There is an apt adage: 'no fat, no leeway'. Finally, be aware of who owns whom: previously old-established, medium-sized contractors now owned by major companies may be under pressure to maximise profit at any cost.

It should be remembered that the continuous employment of one contractor is not necessarily beneficial to the building industry or to buildings as a whole. We should not get unduly 'precious' about conservation work; most traditional (if not specialist) building skills may be used on such projects, at least for the simpler or less important works. Providing such experience for tradesmen helps to educate the building industry, and works towards lessening the amount of insensitive work to buildings, whether listed or not.

Post-contract cost monitoring and control

In contrast to the predictability of new build, complete cost control in conservation is rarely possible, so it is more realistic to talk of cost monitoring. There should be discussion with the cost consultant at the earliest possible stage as to whether proposed works are affordable and the cost-effectiveness of alternative approaches. Once the contract is under way, the quantity surveyor should visit the site regularly and keep records of site operations against the provisional quantities allowed.

When an unforeseen defect is exposed, the quantity surveyor and contractor should be consulted as soon as possible and alternative approaches considered. The professional should never be too proud to ask advice from the tradesmen. The most economic repair method in relation to the importance of the building should be adopted. The contingency sum will be used to cover the cost of the repair, perhaps balanced by savings on provisional sums elsewhere or by postponing certain other repairs. Otherwise, it may be necessary to seek further grant aid or request the owner to provide more money.

When preparing financial reports, the professional consultants should weigh their own predictions against those of the contractor, who should be consulted as to where costs on certain items are heading and, when working to provisional hourly rates, how long it is considered those tasks will take to complete (a useful negotiating tool if the times exceed the predictions).

Contractors specialising in historic building work have in the past been more gentlemanly than general building contractors as there has to be much more give and take. A contractor who feels unfairly treated will be able to retaliate, particularly claiming delays every time another variation in a repair is encountered and there is not an immediate decision. The contractor should receive a fair return, and sometimes the professional has

183

to defend the contractor from a client's unreasonable refusal to pay for what is a legitimate extra.

In short, the building process for the cost consultant in collaboration with the architect is as follows:

1. Define the brief.
2. Prepare the budget with care.
3. Produce a tender document suitable for post-contract cost control.
4. Carefully select the list of tenderers.
5. Keep fully conversant with the extent of the work on site.
6. Maintain the contractor's cooperation through continuous dialogue.

As Sir Bernard Feilden has said:

> The conservation of our historic building demands wise management of resources, sound judgment and a clear sense of proportion . . .[7]

> Efficient cost control systems that favour the conservation of historic buildings are desirable in that they make scarce resources go further. Flexibility, however, is essential, as each historic building is individual . . .[8]

Endnotes

1   Historic Houses Association, *Listed Houses: Incentives for conservation* (HHA, London, 1983).
2   RICS Building Maintenance Information (BMI), Special Report 319: *Review of Rehabilitation Costs* (RICS, London, 2003), p. 1.
3   The Royal Institution of Chartered Surveyors have a register of chartered surveyors accredited in building conservation.
4   Royal Institution of Chartered Surveyors, *SMM7 Standard Method of Measurement of Building Works*, 7th edn (RICS, London, 1988).
5   Ken Davey, *Building Conservation Contracts and Grant Aid: A practical guide* (Spon Press, London, 1992), p. 9.
6   The Joint Contracts Tribunal (JCT), established in 1931, produces contracts, guidance notes and other standard documentation for the construction industry. Since 1998 the Joint Contracts Tribunal has operated as a company limited by guarantee and is responsible for managing the JCT Council.
7   Bernard M. Feilden, *Conservation of Historic Buildings* (Architectural Press, London, 1994), p. vii.
8   *Ibid.*, p. 255.

16 Maintenance in conservation

Nigel Dann and Timothy Cantell

Introduction

Maintenance may be less exciting than a makeover and less glamorous than a heroic rescue after years of inattention, but maintenance is the most sustainable and suitable way to manage historic buildings.

This chapter sets out the advantages of preventative maintenance, then puts it in the context of sustainability, building conservation and policy developments, particularly the conservation management plan. The relationship of maintenance to the principal of minimal intervention is fundamental. The question as to why maintenance is not more widely practised is addressed and best practice among corporate and individual owners is reviewed. Some lessons are learned from European experience and from maintenance inspection services in the UK. In conclusion, a comprehensive national maintenance strategy is called for.

Benefits of maintenance

The longer maintenance is ignored, rejected or postponed, the more the advantages of the approach are lost to building owners and managers. It is worth rehearsing what these advantages are. They lie in five areas.

For the owner, maintenance retains the building's appearance and value and safeguards the investment. Clearing gutters or fixing a slipped tile avoids costly problems later.

In social terms, maintenance reduces the cost and disruption to occupants that flow from building failures and from occasional large-scale restoration. Maintenance makes it more likely that dangers – for example, from loose coping stones or a broken handrail – will be spotted before damage and injury result.

In environmental terms, maintenance means less material is used and consequently it reduces extraction, processing, transport, waste and energy use. It prolongs the use of the embodied energy in the built fabric. It contributes to sustainable development and urban and rural regeneration, and reduces the pressure for new build on greenfield sites.

In cultural terms, maintenance safeguards historic fabric because less material is lost in regular, small-scale repair than in disruptive and extensive restoration. Maintenance is central to protecting cultural significance or value because it is the least destructive of all the 'interventions' which inevitably occur in the process of conserving historic buildings.

In economic terms, maintenance brings business that is steady and counter-cyclical and that particularly boosts small and medium-sized reputable enterprises.

In addition, well-maintained historic buildings improve the quality of life for everyone, help to attract investment to an area, contribute to regeneration and provide a source of local pride and sense of place.

Preventative maintenance

Maintenance is defined here as any activity such as cleaning, painting and minor repair carried out systematically, on a planned cycle and based on regular inspection. Maintenance of historic buildings is most beneficial in conservation terms when it is preventative: that is, intended to reduce or remove the need for repairs, so preventing the loss of fabric which embodies a building's cultural significance.

Preventative maintenance will reduce the probability of decay and the chances that decayed material will have to be renewed. The Society for the Protection of Ancient Buildings advises that such maintenance

> will not only restrain, or even obviate, the need for repairs later, it will prevent the loss of original fabric and is cost-effective. Without such action, owners are often surprised how quickly a structure can deteriorate. Resultant corrective maintenance is disruptive and costly in both fabric and financial terms.[1]

The legislative context

The legislative context for the conservation of historic buildings is the planning system. Planning Policy Statement 1: *Planning and Sustainable Development* (PPS1, 2005)[2] confirmed and reinforced the view, previously developed in a series of other policies, documents and instruments, that the planning system should be based on the principles of sustainability.

There is a clear correlation between the notions of sustainability and the principles of building conservation, in particular the idea of the careful stewardship of finite resources to ensure that the values and benefits we recognise today are passed on to future generations: 'They [the buildings of past times] are not ours. They belong partly to those who built them, and partly to all the generations of mankind who are to follow us.'[3]

Proactive approaches towards the management and maintenance of historic buildings are fundamental to sustainability, and yet are not well supported by the legislative framework for the historic environment. The principal measure, the Planning (Listed Buildings and Conservation Areas) Act 1990, is almost entirely *reactive* in relation to the care of the fabric of listed buildings. It shows concern for maintenance only in so far as it encourages local authority intervention once neglect reaches the point of severe damage.[4] The law is based on the premise that owners have obligations to get permission for any *changes* that affect the historic and architectural character of their building, but they do not have a duty of care regarding the condition of the building. This situation is compounded by extremely limited recourse to any financial aid. In addition there is little supportive advice for owners regarding maintenance other than isolated examples, largely from the more proactive local authorities. In short the message for owners is that maintenance is not very important and that you are 'on your own'.

Planning Policy Guidance Note 15: *Planning and the Historic Environment* (PPG 15, 1994) makes the point clear:

> There is no specific duty on owners to keep their buildings in a good state of repair (though it will normally be in their interests to do so), but local authorities have powers to take action where a historic building has deteriorated to the extent that its preservation may be at risk. [section 7.4][5]

However, the remedies for neglect are cumbersome and potentially expensive for local authorities to administer.

The most significant disincentive to preventative maintenance in current law is the imposition of 17.5% value added tax on repairs and maintenance, while VAT is zero-rated on demolition and alteration work to a listed building when carried out with Listed Building Consent, and on new residential building.

Sustainability, building conservation and policy developments

While the statutory context may not be proactively supportive of maintenance, there have been useful developments in government policy initiatives. These reflect the parallels between sustainability and building conservation and the growing concept of heritage conservation as the 'management of change'. There has been a strong emphasis on the need for the historic environment, whether individual buildings, sites or areas, to be managed more effectively. *Power of Place*,[6] published by the Historic Environment Review Steering Group, emphasised the need to develop processes for managing change. The report also highlighted the importance of methodological approaches which identify cultural significance and its vulnerability, such as conservation plans, as a fundamental prerequisite in appropriate management processes for the built cultural heritage. This was reinforced by the DCMS report, *The Historic Environment: A force for our future*.[7] The growth in the use of management agreements – particularly for twentieth-century listed buildings – and conservation plans can be seen as a response to concern about how change can be managed without devaluing significance. In other words, these address the issue of how cultural significance can evolve while ensuring that existing values are protected, and where possible enhanced.

Perhaps the most coherent development and application of 'new management' thinking towards the historic environment has been the development of conservation plans. This is based on the concepts in the Burra Charter by J.S. Kerr in Australia.[8] In the UK the development of conservation planning has been championed by the Heritage Lottery Fund which insists on development of a conservation management plan prior to grant aid for larger schemes.[9] Arguably the notion of a conservation plan is simply the formalisation of best practice. The methodological approach has some very strong resonance for the management of maintenance and in particular the prioritisation of intervention: the identification and assessment of cultural significance (of the place as a whole and the relative significance of its parts), understanding its vulnerability, and managing it in a way that least damages the identified significance.

Minimal intervention

The identification of cultural significance and its vulnerability, whether or not formalised into a conservation plan or statement, should be the *strategic* starting point for any building conservation activity. There is also a series of approaches to intervention that have gradually become accepted as key building conservation principles: the guiding tactics intended to implement a broader and more long-term strategy. These have been developed over the last 150 years in the UK and can be distilled into five principles:

- **minimal intervention** – doing as little as possible, as much as necessary
- **use of like for like materials** – matching original materials and techniques where possible
- **reversibility** – where possible, adopting repair techniques that are reversible in the future
- **honesty in repair** – not disguising repair as being original
- **the importance of recording** – understanding the structure and what has occurred, and why, before taking any action.

Of these five principles the first, minimal intervention, arguably has the most currency. The idea is that by minimising intervention the material embodying significance will be retained for the future. Maintenance is thus the most appropriate intervention (philosophically and practically) for historic buildings and should therefore take priority over other interventions. The important point about historic buildings, which should affect the approach adopted towards the management of their maintenance, is that the fabric is important in itself – not just because of the function it performs: that is, unlike other buildings, the fabric has cultural significance; the building itself is an artefact. Therefore it is important to emphasise that the terms 'maintenance' and 'repair' should not be used as interchangeably as they might be for non-protected buildings. This is because repair, while it may prolong the life of the element and the building, will also involve damage to the fabric.

The problem with maintenance

Why has maintenance, which has such an apparently important role in building conservation, been understated? The key issues include the following.

- Financial and economic short-termism encourages owners to defer cyclical maintenance in favour of breakdown repairs.
- There is a sense that maintenance provides nothing new for owners – an attitude encouraged by a housing market that seemingly fails to recognise poor maintenance as a significant factor in determining value.
- The costs of supporting maintenance comprise many small long-term sums, which are not easily administered.
- Maintenance has traditionally been a low-status professional and vocational activity, even in building conservation circles, where it does not attract the same kudos as 'major restoration'.
- There had been almost no leadership specifically promoting and developing maintenance in building conservation.

Around 2000, maintenance began to receive more attention. A wide-ranging review of policy for the historic environment, *Power of Place*, recommended that the conservation sector should 'encourage better maintenance'.[10] The government, it continued, should devise and implement a new 'statutory duty of care on owners of listed buildings, scheduled monuments . . . provided it is supported through fiscal incentives and a wider availability of grants'. Public bodies should be accountable 'for their performance in maintaining their historic estates'. For the heritage sector, there must be a 'shift from cure to prevention, by encouraging regular condition surveys and planned maintenance and piloting self-help initiatives', and for the owner, the recommendation was to 'carry out routine maintenance and regular condition surveys'.

The government's response, *The Historic Environment: A force for our future*,[11] accepted the argument and the implication of poor maintenance practice behind it, noting increased awareness of the case for shifting emphasis from 'cure to prevention':

> The Government fully endorses the increasing importance attached to the preventative maintenance of historic fabric. In discussions with English Heritage about future funding priorities, it will explore how a shift of emphasis towards preventative maintenance might be reflected in grant programmes.[12]

Best practice

Best practice in the non-historic sector has been to stress the practical and economic importance of planned maintenance – predicting failure or lack of utility – over the alternative approach of response maintenance, or waiting until there is a problem and then fixing it. Planned maintenance requires far more knowledge of the building stock and requires a range of predictions and monitoring to assess when to intervene. Planned maintenance frequently involves intervention before failure, in the interests of economy – the batching of similar repairs – and avoiding the management hassle related to putting right a failure. The problem with these approaches is that neither is ideal for historic buildings. The overriding tactical principle of conservation, as noted earlier, is that of minimal intervention: planning to intervene *prior* to failure and responding to failures are anathema to this principle. Maintenance planning for historic buildings requires fine judgements if the ideal of minimal intervention is to be attained.

The idea of 'just-in-time' maintenance has begun to gain some currency,[13] and the concept has clear relevance in the context of historic buildings. Here the idea is that intervention should occur when an element or component is about to fail. The implication of this approach is that those managing maintenance require:

- systematic and frequent inspections to inform them as to condition
- the experience and knowledge to make fine judgements about the life cycle and modes of failure concerning the building's elements and components
- information systems that support such an approach.

Inspection becomes a fundamentally important activity; indeed, it becomes the primary planned maintenance activity.

What then characterises best practice for historic buildings? The four key issues are:

- an understanding of the cultural significance of the building as a whole and its components and elements
- regular and informed inspection of the building fabric
- the formulation of an appropriate maintenance policy
- appropriate management and financial resources to implement such a policy.

Many historic buildings are not owned by individuals, but are owned and managed by organisations such as ecclesiastical bodies, educational and commercial organisations, central and local government and agencies, housing associations and, of course, heritage organisations. The majority do not have historic building conservation as a primary aim – unlike organisations such as English Heritage, the National Trust and building preservation trusts. The danger is that such organisations will not fully understand the nature of the historic asset and will have little engagement with appropriate conservation approaches in its maintenance. Table 16.1 summarises best practice in maintenance management, whether for heritage organisations or non-heritage bodies that happen to have listed buildings within their portfolio.[14]

Table 16.1 Best practice in maintenance management.

Maintenance management area	Characteristics of a best practice approach
Corporate objectives and maintenance strategy/policy	• Conservation principles should be the overarching intellectual framework which informs the ethos and implementation of maintenance for listed buildings. Maintenance for such buildings should primarily be concerned with the protection and enhancement of cultural significance, as well as being concerned with continuing utility of the building. • Maintenance management goals and the purpose of the maintenance management function should be explicitly integrated with wider corporate goals.
Management processes, conservation plans and management plans	• Assessments of cultural significance are fundamental to the appropriate management of listed buildings (including their maintenance) and should be implemented through appropriate management plans. The principle of minimal intervention should inform and be reinforced by such management plans.
Programmes and prioritisation	• Maintenance programming should place the emphasis on cyclical preventative maintenance tasks and be driven by the overarching goal of minimal intervention. • Assessments of cultural significance should be central to the prioritisation of maintenance activity.
Condition surveys, inspections and stock data	• A range of inspections at varying frequencies should be carried out. These should be tailored to the significance and vulnerability of the element or material. • Condition surveys should provide an assessment of condition, identify the optimum moment for intervention, and aid the prioritisation of actions and planning for the future.
Information management	• Information on building condition should be stored on an integrated database. It should be easily retrievable and easy to handle for both tactical and strategic purposes. • Systems should be in place which enable information about building condition provided by users other than those directly related to the maintenance department (e.g. visitors) to be incorporated into the maintenance information database.
Financial management and performance measurement	• Budgets should reflect and be informed by the maintenance policy. • A mechanism for feeding back information about maintenance performance to managers and other interested parties should be in place.

Figure 16.1 Maintain our Heritage undertook a pilot maintenance inspection service in the Bath area. High-level areas were accessed in most cases but remote inspection with binoculars was also helpful.

Best practice in organisations

Conclusions from a significant programme of research by Maintain our Heritage (Figure 16.1)[15] and developed by Dann et al.[16] suggested that both heritage and non-heritage organisations fell short of best-practice maintenance, when measured against the criteria in the table (Table 16.1). The areas for potential improvement and the maintenance challenges faced by organisations with listed buildings differed both between heritage and non-heritage organisations, and between the commercial and non-commercial non-heritage organisations.

Heritage organisations were generally clearer about what constitutes good conservation and there was evidence of increasing awareness about the relationship between maintenance and retaining cultural significance. Some of this awareness was being translated into management action by some organisations. However, there was little crossover to heritage organisations of the conceptual and practice advances that have been made in the general maintenance management sector. In particular heritage organisations lacked a systematic and integrated approach to maintenance which links to wider corporate objectives predicated on explicitly using the identification of significance and vulnerability as the key reference point for management decisions and actions. Many also relied on information systems which did not allow for an integrated and strategic approach to the care of their listed buildings.

The heritage organisations generally lacked an explicit set of policies that would provide a framework for prioritising maintenance decisions. It was also clear that little work was done on producing effective management review.

In the non-heritage organisations studied, the approach to maintenance management of the listed stock was driven by attempts to develop efficient processes, rather than a consideration of what the maintenance of listed buildings was attempting to achieve. While many of the organisations had adopted aspects of a best-practice approach to maintenance management, these had not been re-contextualised for the needs of their listed stock.

The lack of understanding of conservation principles was very clear in the commercial non-heritage organisations. For these, the primary value of a listed building related to image. The priority for maintenance activity was, therefore, focused more on retaining the aesthetic appearance of the building and less on a sophisticated assessment of cultural significance. Clearly such emphasis on aesthetics can lead to inappropriate priorities and intervention as it implies that something can be protected by being reproduced.

Best practice and individual owners

The research showed that individual listed-building owners are neither aware of, nor engaged with, conservation principles, adopting an 'if it's not broke, why fix it?' attitude.[17] The majority of owners in a survey involving focus groups, postal questionnaires and telephone interviews prioritised work on their building from a functional rather than a conservation perspective. Moreover, the majority did not translate a sense of obligation to protect the historic nature, listing status or even the 'functional or financial asset' into undertaking preventative maintenance.

While private owners considered that the historical-cultural significance of their buildings was important, they felt that such significance could be conserved by repair and replacement, rather than through regular 'preventative' maintenance and regular inspections for maintenance purposes.

The interviews suggested that owners associated cultural value with aesthetics rather than historic significance. The majority of interviewees saw maintenance and repair as interchangeable concepts. They believed that character can be 'maintained' by extensive repairs or replacement, and where they were concerned about a need to prevent fabric loss the motivation was consideration of cost and/or function.

The majority of survey respondents said that they tried to anticipate the maintenance needs of their building. However, the interviews suggested that in reality this was more of a vague good intent. More significantly, this applied mainly to anticipating when something would need repair or replacement, rather than maintaining it in order to delay failure and the need for repair.

The majority of interviewees did not put aside funds for future maintenance. Although the majority of survey respondents said that some kind of regular inspection of their building was carried out, the interviews suggested that this was rather informal in nature and often carried out either by the interviewees themselves or by a family member or friend. Again, the decision whether to carry out inspections or not seemed to be independent of the listed status of the building.

The research also highlighted that owners are not helped to maintain historic buildings. Advice available to owners about both legal obligations and maintenance and repair was perceived to be poor. Builders were the group from whom advice was most likely to be sought. Advice on maintenance work was not highly valued or sought after, mainly because such work was regarded as consisting of simple and obvious jobs that do not require independent advice or particular expertise.

Learning from other European nations

Dann and Worthing[18] identified four key factors which have led to a more sustainable approach to the maintenance of statutorily protected buildings in various other European nations.

- **A supportive legislative framework.** In the Netherlands and in Belgium there is a duty of care enshrined in law which requires all building owners to undertake a minimum degree of maintenance. This applies to all buildings, not just to historic buildings. In Denmark all owners of apartments (a significant proportion of Danish urban dwellings) have a legally sanctioned duty of care and must maintain the common and shared external elements of the building.
- **An inspection service which enables a less reactive and more planned approach.** The longest-established inspection service for non-ecclesiastical historic buildings is the Monumentenwacht service in the Netherlands. This federal, not-for-profit organisation has been inspecting and reporting on the maintenance priorities of the Netherlands' historic buildings since 1973. Currently the organisation provides prioritised maintenance reports for subscribers representing over 30% of all the listed buildings in the Netherlands. Monumentenwacht have been fundamental in realigning Dutch policy away from repair and towards maintenance for listed buildings. Many attempts have been made to emulate this service in other countries, for example Monumentenwacht in Flanders (established in 1992), Bygningsbevaring in Denmark and Denkmalwacht in Germany (both established in 1999).
- **Financial incentives which encourage a maintenance led approach.** Many European nations have maintenance-focused grants for individual owners of listed buildings; many also have fiscal breaks for owners maintaining a designated and protected building and, where grant aid has been given, there is a best value and sustainability logic to encouraging the maintenance of the asset.
- **Information, support and advice services for owners.** A key issue that encouraged owners of listed buildings in undertaking appropriate

maintenance in the Netherlands, Belgium and Denmark was the extent to which they were positively supported and encouraged to 'do the right thing'. Owners frequently do not have the requisite knowledge or skills to undertake appropriate care of their building. When provided with positive support, particularly in combination with grant aid and tax breaks, the relative importance of maintenance is emphasised.

Maintenance inspection services in the UK

While inspections are the key to preventative maintenance, such inspections are difficult to obtain in the UK market. The construction industry in the main has shown little interest in offering them.

A pilot service set up by Maintain our Heritage in the Bath area showed that it is possible to establish and operate a maintenance inspection service and that such a service could be valued by building owners. The pilot, however, was costly and was made possible by external funding; the market would not have accepted the real cost of providing each inspection. The outcome suggested caution in conceiving of a profitable inspection-only service at a charge that customers will pay. Among possibilities that could be looked at are:

- to combine a maintenance inspection service with related elements such as boiler servicing, drainage protection
- to provide a maintenance inspection service as the centrepiece of a home or building care package
- to offer a maintenance inspection service with the incentive of a discount on buildings insurance
- to provide a maintenance inspection service as a condition of a mortgage or insurance policy.

A maintenance inspection is not a full survey. It aims to identify maintenance issues and set priorities for maintenance work, not to assess the condition of every element of the building. Maintain our Heritage in its pilot offered a service to inspect areas critical for maintenance. The scope of the inspection was thus broadly limited to the external envelope. The rationale was to concentrate on those elements that protect the building from water and damp penetration:

- roof coverings (including flashings to abutments); gutters, downpipes and associated rainwater goods; external wall surfaces and joinery; and drains (Figure 16.2)
- internal roof void
- internal areas where maintenance problems are identified in the external walls and/or joinery
- drains and inspection chambers, by lifting drain covers.[19]

The ten most common problems found in the maintenance inspections undertaken of houses were:

Figure 16.2 Inspection of high-level areas raises health and safety issues. One response is to use roped access, as on this Gloucestershire church.

1. slipped, cracked or missing slates
2. damaged or inadequate flashings
3. blocked gutters or hopper heads
4. leaning or cracked chimney stacks
5. decayed or cracked or open masonry joints
6. loose masonry, particularly at high level
7. cracks or bulges in walls and ceilings
8. rot or insect attack in woodwork, particularly in roof voids and suspended ground floors
9. leaks in water supply and waste pipes
10. blocked low-level ventilators in walls.

There also appears to be scope for a not-for-profit maintenance inspection service targeted at particular sectors such as places of worship, if supported by some level of subsidy.

Moreover, there are other ways of fostering better maintenance practice. The major national heritage bodies could encourage maintenance more proactively, for example, and advice and information about maintenance should be more readily available.

In particular, there is currently no widely available leaflet on maintenance aimed at the general public. Such a leaflet could give:

- a call to action on the benefits of maintenance
- maintenance philosophy/policy in a nutshell
- advice on how to undertake/commission maintenance tasks
- sources of further information and advice.

Conclusion

Maintenance is the most pragmatic and philosophically appropriate intervention, given that conservation seeks to pass on to future generations what we currently identify as being of cultural significance today. If a more effective and sustainable policy for maintenance of the UK's historic buildings is to be instituted, there are a number of issues that require to be addressed:

- Maintenance needs to be promoted as the key intervention for the care of historic buildings at policy level; and owners, both individuals and organisations, need to know that maintenance is of critical importance in sustaining the cultural value that we identify in the fabric of listed buildings.

This needs to have practical expression:

- the development of inspection and other services which support and encourage owners in undertaking appropriate maintenance
- fiscal and other financial mechanisms to reflect and support such an approach.

All of the above requires leadership which supports such developments and integrates policy rather than allowing an incomplete series of piecemeal initiatives. A comprehensive national maintenance strategy is needed that provides wide-ranging official support, advice and encouragement for maintenance. Such a strategy would range from financial assistance, information and advice to an explicit obligation on owners to look after their listed buildings and provide maintenance records in owners' handbooks or sellers' packs.

Endnotes

1   Society for the Protection of Ancient Buildings, Technical Advice Q&A 10: *Preventative Maintenance* (2002). Available from http://www.spab.org.uk/publications_Q&A. html.
2   Office of the Deputy Prime Minister, Planning Policy Statement 1: *Delivering Sustainable Development* (2005). Available from http://www.odpm.gov.uk/stellent/ groups/odpm_control/documents/contentservertemplate/odpm_index. hcst?n=5845&l=3.
3   John Ruskin, *The Lamp of Memory* (Smith Elder, London, 1849).
4   All Acts are available from www.hmso.gov.uk/acts.
5   Department of the Environment, Planning Policy Guidance Note 15: *Planning and the Historic Environment*. (HMSO, London, 1994). Available from http://www.odpm. gov.uk/stellent/groups/odpm_planning/documents/page/odpm_plan_606900.hcsp.
6   English Heritage, *Power of Place: The heritage sector's response to the government's call for a review of policies on the historic environment in England* (English Heritage, London, 2000).
7   Department for Culture, Media and Sport (with Department for Transport, Local Government and the Regions), *The Historic Environment: A force for our future* (HMSO, London, 2001).

8   J. Semple Kerr, *Conservation Plans for Places of European Significance* (National Trust of New South Wales, Sydney, 1996).

9   Heritage Lottery Fund, *Conservation Management Plans: Helping your application*, (London, 2004). Available from www.hlf.org.uk/English/PublicationsAndInfo/ AccessingPublications.

10 English Heritage, *Power of Place*.

11 DoE, *The Historic Environment*.

12 *Ibid.*

13 B. Wood, *Building Care* (Blackwell Publishing, Oxford, 2003).

14 N. Dann and S. Wood, 'Tensions and omissions in maintenance management advice for historic buildings', *Structural Survey* **22**, 3 (2004).

15 University of the West of England, *Best Practice Maintenance Management for Listed Buildings,* report for Maintain our Heritage (University of the West of England, Bristol, 2003). Available at www.maintainourheritage.co.uk.

16 N. Dann, S. Hills, D. Worthing, 'Assessing how organisations approach the maintenance management of listed buildings', *Construction Management and Economics* **24**, 1, 97–104 (2006).

17 University of the West of England, *Individual Owners' Approaches to the Maintenance of their Listed Buildings*, report for Maintain our Heritage (University of the West of England, Bristol, 2003). Available at www.maintainourheritage.co.uk.

18 N. Dann and D. Worthing, *European Approaches to Maintenance of the Historic Built Environment*, International Conference on Conservation Management in the Built Environment, Chartered Institute of Building & University of Westminster, 6–8 September 2005.

19 Maintain our Heritage, *Historic Building Maintenance: A pilot inspection service* (Bath, 2003). Available at www.maintainourheritage.co.uk.

17 Building preservation trusts

Colin Johns

Although conservation legislation in the UK has been effective in controlling demolition of historic buildings and the worst excesses of mutilation, it has been less effective in promoting positive conservation. This does not mean that such activity has not taken place but it has been largely left to private owners and others to take the initiative. Since the mid-1970s at least historic buildings and conservation areas have become highly desirable places to live and substantial private investment has gone into building conservation. This is relatively easy where a building has an obvious future but it is more of a problem where the cost of repair is high in relation to end value or the future of an area is uncertain.

The 1981 working party report on *Britain's Historic Buildings: A policy for their future use*, set up with the encouragement of the Historic Buildings Council for England and the British Tourist Authority, examined the case for the conservation of historic buildings where the original use had been abandoned. Among its many conclusions was the observation that such buildings remained important and valuable assets and efforts should be made to save them.

From time to time we come across buildings in desperate need of repair and probably a new use, and with conservation legislation in place it is easy to ask what the authorities are going to do about it. With the current repairs notice and urgent works procedures this is a valid question, but in years gone by the only option to save a neglected and unloved building was direct action. For some the challenge of rescuing a historic building is so compelling that they will give up their own time and money to do so. For committed conservationists and community groups, one way to achieve this is by setting up a building preservation trust (BPT).

The origins of the trust movement go back to the early part of the twentieth century, with the establishment of the Cambridge Preservation Society in 1929 followed by the Bath Preservation Society in 1934. The Plymouth Barbican Association Ltd was formed in 1957 and the Kings Lynn Preservation Trust in 1959, with a further eight trusts added in the 1960s. By the end of 1979 there were 39 trusts in existence, a number that by the end of 1989 had grown to 92 and by 1998 to 169. The last quarter of the twentieth century was an active time for the trust movement and at the beginning of the twenty-first century there are probably around 300 trusts in existence.

There are two principal types of BPTs: those formed to do more than one project, on the revolving-fund basis, and those formed to save a particular building or site. There are also several trusts set up to save and manage historic property, among the most well known being the Landmark Trust and the Vivat Trust, both of which run successful holiday-letting businesses.

Trusts vary in size, from a small group of like-minded people getting together to tackle a threatened building or buildings in a single town, to a county- or district-wide organisation established at the initiative of the local authority. There are also several country-wide trusts.

Popular support for historic building projects gave substantial impetus to the growth in BPTs, assisted by expanding local and central government conservation policies. The 1969 survey of the work of building preservation trusts was further encouragement. This study, initiated by the Civic Trust with financial support from the Ministry of Housing and Local Government looked at the work of the twenty-one trusts then in existence. Its purpose was:

- to examine the success and defects of existing trusts and revolving funds in Britain and, where appropriate, abroad
- to analyse the basic criteria for success – the order of capital required, the optimum area/population base, and the advantages and disadvantages of different forms of constitution
- to evaluate the basic validity of the trust concept having potentially wider application.

The study group was asked to consider compiling a simple handbook of advice for those willing to start trusts, including a model form of constitution and articles of association. The group was also asked to make proposals for action at national level, covering the United Kingdom as a whole. Special mention was made of successful early projects including the National Trust for Scotland, the Little Houses Improvement Scheme and the rehabilitation of Abbet Street, Faversham. There was also reference to the 1967 requirement for the enhancement of conservation areas.

The revolving-fund idea is often seen to have originated with a 1957 proposal by the National Trust for Scotland for the Fife villages. The idea was formulated by the trust on the basis that if all of the properties that it restored had to be held inalienably, it obviously could not do a very large number, but if it could buy, restore and resell, possible even at a profit in some cases, the money available would obviously go very much further. In 1959 the trust allocated £10000 from its general funds and the scheme was launched, with the first project begun in 1960. By 1971 the trust had succeeded in rescuing fifty-three buildings in the Fife villages. Funding came from the National Trust for Scotland, local authorities and the Pilgrim Trust. The policy of saving vernacular buildings had been central to the trust since its formation in 1931; the innovative idea was the revolving fund. This scheme attracted much interest and was often cited as an example for others to follow.

Among the conclusions and recommendations of the Civic Trust report published in 1971 were suggestions to set up a supporting organisation to provide for an exchange of information, to create a National Buildings Conservation Fund (NBCF) with a target of £1 million with loans from the fund to be made available to local trusts, and that action be taken by central government to establish the NBCF.

In May 1972 the Council of Europe's Committee of Ministers endorsed the concept of European Architectural Heritage Year and defined the objects as follows:

- to awaken the interests of the European peoples in their common architectural heritage
- to protect and enhance buildings and areas of architectural interest
- to conserve the character of old towns and villages
- to assure for ancient buildings a living role in contemporary society.

The Civic Trust was appointed as the UK Secretariat of European Architectural Heritage Year in 1975 to organise a programme of activity for the UK. One of the principal aims of the UK campaign was to establish a national Architectural Heritage Fund (AHF) to provide loan capital to local preservation trusts. The Secretary of State for the Environment promised to match the money raised pound for pound up to a maximum of £500 000, which it was hoped would provide the fund with starting capital of £1 million.

Founded in 1976, the AHF is an independent charity that operates as a national revolving fund across the whole of the UK. Its foremost purpose is to give preservation trusts and other charities access to working capital for projects to rescue and rehabilitate historic buildings. The AHF makes low-interest loans to organisations with charitable status for projects involving historic buildings that are listed or in a conservation area and in need of repair and rehabilitation, and where there is a change of ownership or of use. From its initial £1 million target in 1975, the AHF now has resources in excess of £13 million.

In addition, AHF can offer Options Appraisal Grants, Project Administration Grants, Project Organiser Grants, Refundable Project Development Grants and Refundable Working Capital Grants. Over the first twenty years of its life the AHF contracted 330 low-interest loans amounting to over £22.5 million. AHF produces an illustrated annual review giving details of projects that have been considered or completed during the year.

The other noteworthy event in 1975 was the founding of SAVE Britain's Heritage, commonly known as SAVE. As well as campaigning vigorously for the restoration of historic buildings, this organisation set out to show that almost any historic building could be rescued given sufficient determination and imagination. SAVE later went on to take up the direct challenge of acquiring and repairing historic buildings, at the same time arguing the economic case for their conservation.

Revolving-fund and single-project BPTs are voluntary organisations but their business involves buying, repairing and selling or managing property: in effect, property development. For this it is necessary to mobilise large amounts of money and, in order to protect those who take decisions from being personally responsible, the activity needs a separate legal existence.

The most appropriate way of giving an organisation this identity is incorporation as a company limited by guarantee, usually with charitable status. The principal advantage of forming a company limited by guarantee is that, provided due care is exercised in the administration, the liability of individual members is limited. An incorporated BPT has clearly defined powers that facilitate its business operations.

The first requirement when setting up a revolving-fund BPT is to decide what the trust wants to do – in other words, to define its objectives. To achieve charitable status, all of the BPT's objectives must be charitable and comply with charity law. To assist the process of registration comprehensive advice is available from the Charity Commission. For a BPT the essential ingredient for charitable status is 'the preservation of buildings of architectural or historic interest for the benefit of the people of a particular town or county'. The trust must also define its geographical limit of operations.

The articles of association of a trust set out the terms of reference of the governing body. Where the trust is a company limited by guarantee, the trustees are in effect company directors and must comply with the relevant clauses of the Companies Act. The AHF produces an information pack, including a 'standard governing document' on the setting up of a trust.

In the past many trusts have been established for the purpose of saving a particular building and this applies especially where the building is to be brought into community use. For revolving-fund trusts the situation is different, in that the exercise is primarily concerned with saving buildings and returning them to the property market.

Building preservation trusts can be extremely effective when working with local authorities, both in the identification of buildings at risk and in securing their future. There are a number of examples where a local trust has provided the essential back-up by the use of repairs notices, and the obvious way for the two organisations to proceed is for the local authority to acquire the property, possibly by compulsory purchase, and immediately to pass it to the trust for renovation and resale.

In the latter half of the twentieth century many trust projects were funded by the relevant government departments or by the Historic Buildings Council for England. Substantial contributions were made by the Department of the Environment and later by English Heritage. More recently the key benefactor has been the Heritage Lottery Fund. The links between conservation and regeneration, identified some years ago by SAVE, continue to influence the policies of funding agencies.

There has thus been a substantial move away from conservation projects for their own sake and towards projects that make a contribution to the regeneration of an area by improving the environment and providing renovated buildings for residential or employment use. The potential for such activity was recognised in the 2000 English Heritage review, *Power of Place*, and in the 2001 government response, *The Historic Environment: A force for our future*. The theme of the review was that conservation-led regeneration is successful because places matter to people, and that the retention of buildings can also mean the retention and improvement of communities.

It seems likely that a number of building preservation trusts will focus their attention on regeneration potential, particularly where this can unlock funds not previously available. One of the conclusions of the 2004 report by the then Office of the Deputy Prime Minister (ODPM) Housing, Planning, Local Government and the Regions Committee on the Role of Historic Buildings in Urban Regeneration was that building preservation trusts perform an important role in bringing back into use neglected buildings which the private sector is not interested in; this has certainly been demonstrated by the many successful schemes.

In the 1971 Civic Trust report mention was made of the desirability of creating a second-tier supporting organisation to provide for an exchange of information between BPTs. Encouraged by the AHF and following the Second AHF Conference in 1988, the UK Association of Preservation Trusts was formed in 1989. APT is an independent charity formed 'to encourage and assist BPTs, both individually and collectively, to expand their capacity to preserve the built heritage'. It arranges conferences and seminars and has produced a series of guidance notes for trusts.

Over the decades the Architectural Heritage Fund annual reviews collectively illustrate a large number of remarkable rescue projects of all shapes and sizes. The buildings range from small dwellings to redundant churches and public buildings, many of which, in addition to their significant architectural and historic interest, provide attractive new uses and contribute to the regeneration of the areas in which they stand. This is the most significant initiative whereby building preservation trusts take risks that the private sector is unwilling to embrace. Establishing confidence is often the first step to effective regeneration.

Few buildings are demolished simply because they are in such poor condition that they cannot be saved. They are usually demolished because the cost of repair exceeds the value on completion or because the site is more valuable without the building. There may also be social or political reasons encouraging demolition. These may be related to a replacement building elsewhere or simply changing economic forces. If there is no demand for a building in its present form it is likely to be neglected. In some cases it has also been argued that restrictive conservation policies, for example on alterations or change of use, have contributed to the problem. Building preservation trusts can bring in an entirely new agenda and provide solutions not found elsewhere, and this is why the potential sale of a building, possibly to a trust, is a material consideration in all formal applications for demolition. Trusts can create opportunities not obtainable within the private or public sectors and thereby add to the quality of life in the areas in which they operate. The concept remains valid and there is more that can be done.

18 Valuing our heritage

David H. Tomback

Introduction

Some people may ask why we bother to value our heritage at all: surely our heritage is beyond monetary value? But life is not as simple as that – the built heritage does have to be valued, and the methods by which society places value, and the accuracy of these methods, have over the past twenty years become increasingly significant when decisions that affect the future of our historic environment are being considered.

The concept of valuation is well established, and the Royal Institution of Chartered Surveyors (RICS) provides rules and regulations which guide, inform and steer how valuations are carried out for a large variety of purposes. The difference between 'value' and 'worth', different methodologies and the definitions attached to assumptions are all set out in the RICS Appraisal and Valuation Manual (Red Book). When valuing our heritage, however, special and unique factors can often come into play, and how one measures non-financial elements such as historical importance also needs to be considered. This short chapter raises some of the principal issues and highlights areas for further research.

What is heritage?

The perception of heritage varies from individual to individual. The government, in Planning Policy Guidance Note 15 (PPG15), states from the outset:

> It is fundamental to the Government's policies for environmental stewardship that there should be effective protection of all aspects of our historic environment. The physical survivals of our past are to be valued and protected for their own sake, as a central part of our cultural heritage and our sense of national identity.[1]

Indeed the value of historic buildings and places is confirmed by the United Kingdom government's being a signatory to various international conventions (see Chapter 4).

As Rypkema notes:

> Preservationists often talk about the 'value' of historic properties: the social value, cultural value, aesthetic value, urban context value, architectural value,

historical value and sense of place. In fact, one of the strongest arguments for preservation ought to be that a historic building has multiple layers of 'value' to its community.[2]

These values are difficult to quantify. However, Tiesdell *et al.* suggest that 'economic value' should underpin such justification for preservation:

> The desire to preserve must ultimately be a rational economic and commercial choice; problems will arise where buildings are preserved only as a consequence of legal and land use planning controls.[3]

Our historic environment contains a vast range of structures from stone circles, castles, churches and cathedrals to classical Georgian town and country houses, through to industrial properties such as mines, factories, gasholders and mills. In addition, there are historic landscapes. In fact it is easy to come to the conclusion that everything has (or will have) heritage value – if not now, then in the future. It is a myth that heritage automatically equals beauty but, as we shall see later, beauty – or at least visual architectural or historic quality – can have an important role in valuation.

How do we value?

The old adage is that valuation is an art, not a science. In truth, it is a combination of the two. A MORI poll in 2003 said:

> there is no doubt that society as a whole values heritage. Market research surveys consistently reveal that people associate a host of positive values with the historic environment around them. More than 80% of people in different parts of the country agreed with the statement that 'the heritage in my local area is worth saving'. Nine out of ten people agreed that their local area counted as much as 'heritage' as castles and country houses, while 82% agreed that heritage was 'fun'; 98% of respondents thought that heritage was important to teach children about; 95% thought that heritage was important for giving us places to visit and things to see and do; 94% thought that heritage encouraged tourists to visit; 88% thought that heritage created jobs and boosted the economy and 75% agreed that their lives were richer for having the opportunity to visit and see examples of heritage.[4]

English Heritage's strategy for encouraging people to develop a positive regard for heritage is based around the **Heritage Circle** – the idea of a cycle of understanding, valuing, caring and enjoying (Figure 18.1).[5]

In the case of Greenside, Wentworth, a Grade II listed Modern Movement residence of 1937 by Connell, Ward and Lucas illegally demolished by its owner, part of the argument for demolition was that the building was 'ugly'. This is an extreme example of what can happen when a heritage building is not valued – in this case by the owner, who now has a criminal record.

Historic buildings are usually valued using traditional valuation methods; they have a market value and the valuations are carried out by chartered surveyors, or purchasers, in the normal way. Well, not quite – in some cases

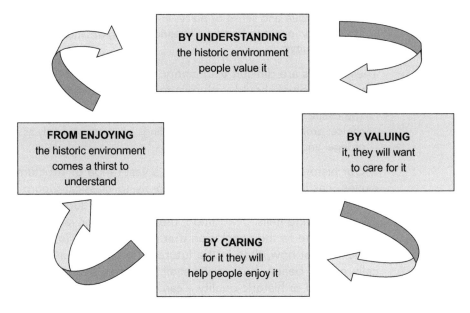

Figure 18.1 The heritage cycle.

the 'love factor' comes into play and purchasers can, and do, pay above the market value . . . but more on that later.

The valuation methodologies can be split into traditional market valuation methods, which are used for the majority of historic 'heritage' buildings, and the non-market valuation methods, which attempt to place a value on the non-financial benefits that the non-functional heritage building or monument brings to society.

Market value valuation methods

The two most common and easily understandable methods are the comparable and investment methods. Put simply, the **comparable method** involves analysing recent transactions of similar properties in the same location and applying a rate per square metre to the property to be valued, having made adjustments for location, condition and so forth. Used for commercial and residential valuations this is an accepted and reasonably accurate method, but when applied to historic houses things can be more complex. This is because, with some exceptions such as properties in the Royal Crescent in Bath, no two historic houses are identical and the intangible elements of architectural style, attractiveness and indeed history, as well as repair liability and possible higher maintenance, all complicate the approach.

The **investment method** is where an income stream (or potential income stream) is capitalised at a yield determined by the market. Careful analysis of comparable transactions is required to judge the income flow projection

The figure contains the following text in boxes:

BY UNDERSTANDING the historic environment people value it

BY VALUING it, they will want to care for it

BY CARING for it they will help people enjoy it

FROM ENJOYING the historic environment comes a thirst to understand

– that is, if there is likely to be a rent review in the future, what is the open market rental and the correct yield? As the investment method applies mainly to commercial properties such as offices, shops, factories and warehouses, commercial reality steps in and the capital value can be accurately calculated, except where an owner has a particular desire to occupy a specific property.

Another accepted valuation technique which applies to businesses, especially hotels, is the **going concern** approach. Some historic hotels and large country houses converted to hotels can generate a 'heritage premium', and a higher room rental can be achieved because of the historical ambience and architectural style of the property.

There are also other valuation methods such as the depreciated replacement cost method, but these are beyond the present scope.

Non-market value valuation methods

The most difficult question for society is how does one value the intangible? Economists have over a long period of time come up with several methodologies. In *The Value of Conservation?: A literature review of the economic and social value of the cultural built heritage*, published in 1996 by the then Department of National Heritage, English Heritage and the Royal Institution of Chartered Surveyors, the different methodologies were set out.[6] The main methods are as follows.

The **contingent valuation method (CVM)** directly questions consumers on their stated willingness to pay for, say, an environmental improvement, or their willingness to accept compensation for a fall in the quality of the environment. There are various applications and formats that CVM can take but respondents might, for example, be questioned as to how much they would be willing to pay to improve the setting of Stonehenge.

The **hedonic pricing method (HPM)** is similar to the traditional comparable method and was developed by Rosen (1974).[7] It aims to determine the relationship between the attributes of a good and its price and is arguably the most theoretically rigorous of the valuation methods. A large number of hedonic studies considering the effect of environmental and neighbourhood variables on house prices have been undertaken in the past and there is a significant body of research into the impact of architectural style and historic zone designation on property valuation. The basis is that any differentiated product unit can be viewed as a bundle of characteristics, each with its own implicit (or shadow) price. In the case of housing, for example, the characteristics may be structural, such as number of bedrooms, size of plot or presence or absence of a garage, and can range through to environmental matters such as air quality, the presence of views, noise levels and even crime rate.

The **travel cost method (TCM)**, developed by Clawson and Knetsch (1966),[8] is a simpler methodology than HPM in that it is based on the premise that the cost of travel to recreational sites can be used as a measure of visitors' willingness to pay.

The value to society of these methods is that important decisions involving our heritage can be better informed and the non-financial benefit to society can be included in the decision-making process. The classic example was the decision, in principle, to build the tunnel at Stonehenge, where the CVM was used and the Treasury was persuaded to accept the approach as a valid method. Evidence from contingent valuation studies was an important part of English Heritage's case at the 2004 public inquiry in respect of the proposed tunnel.

Another system which helps to assess the 'value' placed on construction projects was developed by Loughborough University. The system, called **Managing Value Delivery in Design (VALiD)**, is designed to help stakeholders understand one another during team formation and provide a comprehensive view on value.[9]

The heritage love factor

In certain situations, particularly in connection with listed residential properties, the end value of a property may be exceeded by the cost of acquisition, repair and conversion cost, which would imply that the purchaser in theory paid too much for the property. Why? The answer may be an underestimation of costs or end value. However, some owners simply fall in love with a property because of its location or, more often, the unique living space that they can envisage, as well as in some cases the history of the property. Classic examples are redundant windmills, Martello towers and water towers.

We might describe this as the 'heritage love factor'. Actually quantifying the heritage love factor is probably impossible, beauty indeed being in the eye of the beholder – just as one collector will pay more than another for an antique object.

The Nationwide Building Society did, however, carry out a survey in 2003 in which they found that certain properties had a premium attached to them according to the year of construction. For example, while buyers paid an 18% premium for houses built between 1714 and 1837 and an 8% premium for Victorian houses, houses built in the 1960s and 1970s were 2% to 3% below the base. This raises the interesting question as to whether a Grade I house is worth more than a Grade II house.

In reality it is probably not possible to compare the value of properties by listing alone, for while their importance can be differentiated, market value, particularly of residential properties, is far more difficult to compare; very few listed buildings are identical and other factors, in particular location, are predominant. When considering cultural value, however, it could be argued that as there are only about 8000 entries at Grade I in England and some 415 000 entries at Grade II, Grade I is therefore scarcer and logically Grade I buildings should be more valuable. Grade I and Grade II* buildings certainly benefit from a higher level of statutory protection; the opportunity for change is inevitably more limited and internal space often less adaptable. This, one would think, would depress value. Nevertheless, top-of-the-market property advertisements make clear that, despite restrictions, listed houses fetch considerable prices, and there is no doubt scope for further research into why this is.

We will not here explore the concept of 'public value'.[10] Economists can understand value as expressed by markets and values expressed in contingent valuation surveys, but have more difficulty with public value that has no obvious cardinal, ordinal or descriptive scales.

Other uses for valuations

The market value[11] of a property is used by local planning authorities and English Heritage when considering applications for enabling development. Enabling development is development which is contrary to the established planning policy and is proposed to secure the future of historic assets. Here, an amended form of development appraisal is used to determine the minimal amount of enabling development.[12]

Both English Heritage and the Heritage Lottery Fund use market valuations when assessing certain types of grant assistance, as do lending bodies when valuing heritage properties for lending purposes.[13]

Conclusion

There is no doubt that the historic environment has immense value to society as it plays a pivotal role in our quality of life and the economic well-being of the country. Increasingly, more sophisticated valuation techniques are being used to measure the non-financial benefits of our heritage to enable government to make sometimes difficult decisions on how to allocate funds for major regeneration projects that affect the historic environment. On a smaller scale, individual heritage properties are usually valued by more conventional methods. It is the duty of the professional valuer to carry out valuations in a proper, considered and, most importantly, accurate manner. Valuations play a vital part in the process of bringing long-term beneficial use back to redundant historic buildings, and understanding of the valuation process is essential for all those involved with our historic environment. There is scope for further guidance and education in this field.

The techniques referred to above to evaluate the non-financial benefits of the heritage, in particular the contingent valuation method, will play a greater role in decision-making in respect of large – and often controversial – development schemes that affect the historic environment and our heritage.

It is inevitable that society will move towards greater energy conservation and sustainability, and local planning authorities will in future be made to look more closely at the reuse of historic buildings before allowing a new building to be erected. The chartered surveyor and the economist will increasingly collaborate in the complex area of considering how society values its heritage. For those involved with the historic environment, a greater understanding of valuation issues and a merging of traditional and non-traditional methods will lead to better decision-making for the benefit of future generations.

Endnotes

1 PPG15, p. 1 (1.1).
2 Donovan D. Rypkema, *The Economics of Historic Preservation: A community leader's guide* (National Trust for Historic Preservation, Washington, 1994), p. 206.
3 Steven Tiesdell, Taner Oc and Tim Heath, *Revitalizing Historic Urban Quarters* (Architectural Press, London, 1996), p. 11.
4 MORI poll for English Heritage (2003).
5 English Heritage, *Making the Past Part of our Future: English Heritage strategy 2005–2010* (English Heritage, London, 2005), p. 3.
6 Gerald Allison, *The Value of Conservation?: A literature review of the economic and social value of the cultural built heritage* (English Heritage, London, 1996).
7 S. Rosen, 'Hedonic prices and implicit markets', *Journal of Political Economy* **82** (1974), 34–55. See also J. Nelson, 'Residential choice, hedonic prices, and the demand for urban air quality', *Journal of Urban Economics* **5** (1978), 357–69.
8 Jack Clawson and Marion Knetsch, *Economics of Outdoor Recreation* (Johns Hopkins Press, Baltimore, 1966).
9 Derek S. Thomson, Simon A. Austin, Hannah Devine-Wright and Grant R. Mills, 'Managing value and quality in design', *Building Research and Information* **31**, 5 (2003), 334–45, www.valueindesign.com.
10 Robert Hewison and John Holden, *Challenge and Change: HLF and cultural value* (London, 2005), is a report to the Heritage Lottery Fund by Demos (The Every Day Democracy Thinktank) on the concept of public value. It puts forward the model of the 'cultural value triangle' and the recognition of intrinsic values, instrumental values and institutional values.
11 Royal Institution of Chartered Surveyors, *RICS Appraisal and Valuation Standards 5th Edition* (the Red Book), www.ricsbooks.com, is the principal guidance document which sets out how valuation surveyors should approach and carry out valuations.
12 English Heritage, *Enabling Development and the Conservation of Heritage Assets: Policy statement and practical guide to assessment* (English Heritage, London, 2001).
13 Stephen Boniface, 'Mortgage valuations on historic buildings', in *The Building Conservation Directory* (Cathedral Communications, Tisbury, 1998).

Index